ADVENTURES AND MISADVENTURES

COUNTRY VET STORIES AND MORE

RUSSEL HUNTER DVM

This book is dedicated to my grandfather, Simon Konnerth, who traveled half way around the world to settle here.

Tillasana Publishers

42493 Hillcrest Loop

Astoria, Oregon 97103

rlhdvm@gmail.com

❀ Formatted with Vellum

ACKNOWLEDGMENTS

I have been assisted by some wonderful friends with extraordinary literary talents.

Thank you Julie Brown, Barbara Hansel, and Mandy Schimelpfenig for your thoughtful ideas and encouragement.

Thank you Jan Johnson for guiding me from taking the written word and moving it into a book.

And finally, thank you Molly, my wife, and a good friend, Marla Brownlee, for carefully reading my stories. It's something I could never do.

INTRODUCTION

It turns out that I have more country vet stories. Stories about my life on farms and ranches in the country. About the people, their animals, maybe their lives too. Certainly, about my life.

It also turns out I have another group of stories. Those stories are outside my life as a country vet.

I love to travel, see the world close and far, climb mountains, kayak wild rivers and explore places rarely traveled.

These were often wonderful, and arduous adventures that became lifelong memories.

But sometimes, things didn't quite work out as planned. These became lifelong memories too, but for a different reason. These were misadventures. Fortunately, nobody got hurt or maimed. There were no fatalities. There were certainly some frightening moments. Nobody ended up arrested nor jailed. Although that came pretty close to happening a few times.

These stories are all true. Maybe some names have changed and a few details too. But they really happened.

1

———

t The Mallory

THIS WAS AN OLDER, long-established hotel in Portland. It had a certain class about it you don't find in newer hotels. More private, quiet, good food, really good food, and easy walking distance to much of the downtown.

It seems like a lot of newer hotels, the ones you can afford, are like Motel 6's stacked up for multiple stories. And the free breakfast is a frozen waffle to thaw in a microwave. Afterwards, it tastes and chews like cardboard. The coffee pot is always on, too. It seems like the same coffee for days. But it will keep you awake with a distressed stomach for the rest of the day, if not more.

That was not the Mallory.

It was our getaway for years. Especially after the kids were gone. That wasn't the real reason, though. The real reason was our old house.

The house came with the farm, a small river, and a wonderful barn that became my vet office and clinic. We had pasture and forest too.

But the house was sort of an add-on, like free kittens or zucchini squash you find along the road. It was built in the very late 1890s. We knew this because the only closet was papered with the Oregonian newspaper from 1897.

By the way, according to that newspaper, the coal workers were on strike at the time. They wanted $300 a month in wages. I'm not sure of their working conditions or length of working day or week. Probably not ideal by any standards. Regardless, that didn't seem like bad money for that long ago.

It was a board house, which means it didn't have a frame made of 2x4s or 2x6s. The sides consisted of rough-cut 2x12s that ran upright to the roof. Part of the house had two stories, so pretty high in those places. The outside was cedar planking and the inside was, well, mixed panels of wood. No insulation possible and winter could be a challenge. We used wood heat and as the winter got colder, all of us sat closer and closer to the stove. Humans and animals.

All the plumbing had to be on the outside, too, since it couldn't be put in the walls. That meant we had to keep the water running all night in really cold weather.

Now you're starting to see why the Mallory looked so good.

There was no foundation either. It was set on river stones, the biggest that could be carried from the nearby river. Naturally, things had settled over the years and the house wasn't level.

We did some upgrades, like a foundation, double-paned windows, and attic insulation. I tried floor insulation under the house, but it was too claustrophobic, filled with dead creatures, snakes and spiders.

I cut more wood instead.

And still, the wind could blow through cracks we didn't see or could ever find.

Now, it was time for a romantic weekend at the Mallory. Time to get away from the vet practice and the house I was just talking about. I made reservations and off we went. And a delightful weekend it was, enjoying downtown Portland and the comforts of the Mallory.

Sunday morning. We were just lying around in bed after a romantic night. Quite pleasant. Peaceful too. Finally, Molly got up to take a shower.

I was awake now and thinking what to do next. Then I remembered. It's Sunday morning, early, too. And the Sunday Oregonian would be waiting just outside the door. All I'd have to do is crack the door, reach out and grab it. No clothing required. Just be fast.

I went to the door. I cracked it open, but the newspaper was farther out than I expected. Holding the door, I reached for it. It was too far yet. I stretched more. I still couldn't reach it. I stretched even more, so much that I let go of the door. Just as I grabbed the newspaper, the door closed behind me. It locked.

Of course, I tried. But it was locked and I knew that.

Instantly, I knew what a convict experiences when their cell is closed and locked.

I was trapped naked in a major hotel in Portland. With a newspaper in hand.

A lot of things can go through your head almost instantly. Mainly, somebody will see me, call security, I'll be arrested for indecency and hauled off to jail. My wife will not even know what happened to me.

Nobody was around. Too early, I guess. Or maybe just lucky.

In total panic, I turned to the door and started beating on it and yelling for my wife. If she had decided to take a long shower, well, the jail possibility was real.

Finally, she cracked the door open. It was a perfect time for extortion, but all she could do was laugh and let me in.

2

———

C omfort Suite

JENNIFER LIVED on a small farm not far from me. Her dad worked at the mill. Her mom raised goats and an occasional orphan fawn.

She had a horse, Spook. Part Morgan, part Thoroughbred. It's safe to say they loved each other. She rode Spook all over the local hills. Her time with him and her experiences with him led her to a career caring for the city of Portland's police horses.

Spook had an interesting habit of resting his head on her shoulders. Quite endearing. And, as it turns out, pretty important for a future situation.

One day after riding, Jennifer led Spook to his pasture and turned him loose.

Instead of removing his halter, she just unclipped the lead rope and took it back to the barn.

Some time later, probably several hours, she went back to bring him in for the night.

During that time, he had trapped a back leg hoof in the halter. Sometimes that can happen with horses. He struggled and struggled to get it out. Now he was down, exhausted, and with severe damage to his neck muscles.

That's how Jennifer found him. She had to cut the halter off. He could get up, but could barely use his neck. His head hung almost to ground level because the neck muscles were very damaged. When that happens to a horse, their head swells up. Nostrils too. And that could be life threatening.

She called me and I arrived as soon as I could. His head swelling was increasing and looking more and more dangerous. I treated him with drugs to reduce the swelling and pain. But I knew we needed to do something with his head.

She led him to his stall in an open-sided barn. We concocted a sling from an extra saddle cinch and put it at normal head level. Then we put water on one side and feed on the other side. At the same height as the sling.

Jennifer led him to the sling. We put his head in it and he rested. I'm not sure how many horses would do that. He seemed perfectly content. That's when Jennifer told me about his habit of resting his head on her shoulder.

I visited him daily for some time after that to evaluate him. He quickly learned to grab a bite of food or sip of water and return to his sling.

Spook made a fast recovery. A total recovery.

Much later, Jennifer told me he used the sling for sleeping the rest of his life.

3

B ad Eggs

"YOU NEED TO HELP ME. I have something to show you," she spoke to me in her eastern European accent on the telephone.

"Sure, no problem," I answered.

"What's the problem?" I then asked.

"I have to show you," was her reply.

Over the years I have learned some people are uncomfortable talking about certain things. Especially sexual things. Like reproductive anatomy.

I set a time for the next morning for her visit.

Betty was from Hungary. And moved here years ago with her new husband. He had been in the military in Europe. I think that sort of explained how they met up.

Most Eastern Europeans who come from small towns or rural areas have farming backgrounds.

I know this because my grandfather came from northern Romania, the Transylvania part. It borders with Hungary and was actually part of Hungary before WWI.

He grew up in a small village where almost everybody farmed. They had vegetables, fruit trees, chickens, sheep, cattle and horses.

Every morning, they would send their cattle and horses into the main street where a gypsy family would herd them to community grazing grounds. In the evenings, they would herd them back for the night. The sheep stayed with herders and their dogs in the mountains.

In the end, everybody grew up learning how to farm and care for their animals. Betty did too.

Betty showed up the next morning with her son Frankie. She was carrying a large brown paper bag.

I didn't say anything, I just waited. She hesitated.

Finally, Frankie said, "Show him what's in the bag."

She opened it, it seemed with reluctance. I looked in.

Inside were three young chickens. They all had very obvious birth defects. Crooked bills, crooked necks, crooked legs and feet.

The young chickens were simply freaks with no hope for a normal life.

Betty was still quiet. I was confused. Finally, Frankie told me the story.

His mother had recently traveled to eastern Europe to her home village.

This is something she had done over the years. Lots of memories, lots of friends, lots of family still there. It was a natural thing to do.

This time she realized how much she missed the chickens the family always had. The hens, the rooster, the chicks and especially the eggs.

So, she concocted a plan. She smuggled about a dozen fertile eggs to bring home. I'm not sure how she did that, carry-on or regular baggage. Life was simpler then and you could get away with more.

Anyway, she made it home with the eggs intact, put them in her incubator and waited. Not many hatched; after all, it wasn't an ideal way to handle fertilized eggs. Flying them halfway around the world just wrapped up and packed away.

Three hatched. The three she brought to me. They all had terrible birth defects. Probably something about their journey caused it. I'm not sure exactly what, but they were not going to have a normal life. I told her that. Frankie knew that. I think Betty did too, but she was hoping for a miracle.

I had no miracle and the only humane thing to do was to mercifully end their lives. Which I did.

4

———————

Calling In The Law

I HAD KNOWN her for some time. Phylis. She had horses that I occasionally cared for. On and off over the years. Strong opinions, judgmental and temperamental. More than I realized as it turned out. A lot more.

I was working the south end of our county making rounds. I would collect a bunch of calls and travel to that area once a week. Usually on Wednesdays. People knew I was coming and would contact me ahead of time. It saved me a lot of travel time and them some money.

It was Wednesday and Phylis had asked me to stop and look at her horse. Apparently, it had injured several legs.

The horse was in a small field with makeshift shelter for bad weather. The grass was grazed down to that of a regularly mowed front lawn. A single barbed wire fence surrounded it. It looked like a second or third grade class project. The wire looped from small post to small post at knee level. Sometimes even lower.

I couldn't believe the horse stayed in. But it was older and had a very quiet personality.

She was concerned about a large number of apparent scratches on its front legs. I examined the legs. Although the injuries were numerous, they were largely superficial. I looked around for the cause. It became apparent, to me at least, that the horse was stepping over this questionable fence to find more grass on the other side.

I explained my findings and thoughts.

She didn't take my ideas well. Not at all.

She told me, in no uncertain terms and with increasing hostility, what was going on. In her opinion anyway. The neighbor's dogs were doing it.

As it turns out, the neighbor had two terrier-sized dogs not too far away. I could see them. They didn't seem too threatening to any horse from my standpoint.

She disagreed and was now getting pretty excited as our discussion went on. Furthermore, she was convinced the two small dogs trapped the horse in its shelter at night and attacked its legs.

That didn't seem likely or remotely possible to me. Horses can defend themselves in different ways, especially from small dogs. Biting, kicking, running. Anyway, dogs are not that stupid to attack a horse. Especially little dogs.

That's not to say horses haven't been seriously injured by dogs. Especially large dogs and more than one. I have certainly cared for these injuries over the years, but they are pretty unusual.

My thinking was going nowhere with her. She was getting more irate. Finally, she produced a gun. Then she told me her plan. It was to sit in her car for the night.

When the dogs arrived to attack her horse, she would shoot them.

A very mad woman, with a bizarre idea and now a gun. It was time for me to leave and maybe get help. This was not in my contract, as they say.

It's funny how a gun suddenly appearing changes your behavior. I became very agreeable and left as soon as possible.

A deputy sheriff lived about five miles down the road. This was his territory. I just hoped he might be home. As it turned out, he was. Taking a lunch break.

He found it as alarming as I did and agreed to visit her.

5

———

B^{ud}

I'M LEANING against an old wooden corral. It's been there forever. It's made out of redwood, so it's going to last a lot longer than forever. Standing next to me is Bud. He's dressed as normal. A cowboy hat, wool jacket, blue jeans, and slip-on logger slippers. He's not prepared for action. That's for sure.

In the corral is Jay, a college student who works at our clinic. (Eventually, Jay will finish college and become a veterinarian.) With him is a 5-year-old unbroke horse. Hanging on the corral are a saddle, saddle blanket, a halter with lead rope and a bosal. Bud is going to guide Jay through catching, quieting, saddling, mounting and riding this horse. We have 45 minutes before getting back to the clinic and work. This is our lunch hour.

I know this horse is unbroke. About a year before, its

owner had caught it up, with its mate, in this same corral. The horse was very lame, almost three-legged, with an infection in a front foot. Fortunately, I had a good helper with me. I roped the horse. Before it choked itself fighting, I injected enough sedative to quiet it and then administered a short-acting general anesthetic.

That happened years ago. Our drugs today are much more effective and safer now. But things worked out just fine.

I found the infection, drained it and gave the horse antibiotics. This was going to be a one-time treatment. It worked; the horse was normal now.

Except, still unbroke.

Quietly, Bud talked to Jay. Guiding him through talking to the horse, making it comfortable, finally catching it, putting on a halter. Then the blanket. Then the saddle and finally the bosal on the head for control. Finally, Jay got on and moved the horse about the corral using the bosal to guide the horse's direction.

Truthfully, I don't know how all that happened. It was like magic as far as I was concerned. Not only that, we made it back to work on time.

Actually, it happened because of Bud. Bud Williams. Stop reading right now, go to your computer or phone, and look him up. On Google. Bud Williams Stockmanship.

Looking back, I consider myself lucky, once-in-a-lifetime lucky, to have spent time with Bud. Bud's wonderful wife, Eunice, managed our clinic. Bud often stopped by. Maybe it was because he didn't have a lot else to do. I think he logged some on the side. Worked in the woods, as they say.

Bud had grown up in southern Oregon around animals in the early 1930s. Tall, lean, quiet, but never missing anything. Searching eyes, always observing. Always learning about animals, especially horses and cattle. Today, he would be called a horse whisperer. But he worked with cattle too, a lot. At one point, even reindeer in Alaska. Eventually, he settled in the Midwest and set up his livestock handling school. Stockmanship, it was called. His work and his knowledge became well known the world over. It still is.

Because he had the time, so it seemed, he would offer to go with me on cattle calls. Always herds of cattle.

Sometimes we needed to work cattle into a chute for me to do something. Something like vaccinating for Brucellosis. This was something vets had to do. Because, one, it was a dangerous vaccine to humans. Two, it was required by the government and had to be certified by a vet. This was part of a years' long eradication process because the disease is contagious to cattle and humans.

The ranchers were glad for Bud's help. They let him move the cattle to the chute. Usually without a sound. Certainly not with any force. He just knew how and when to apply slight pressure to move them forward.

This knowledge was part of the teachings at his livestock school.

I was never sure why Bud took me under his wing, so to say. He certainly wasn't bored in his life, although he did hang around a lot.

I think he looked at me as a project. A new, very naive vet, whom he could educate. He was certainly right about that. What I learned from him has stayed with me forever.

Thanks, Bud.

6

____

Deadman's Hole

OUR GRANDSON WAS home from college for the summer. Mack. He was having dinner with us. Then, he and I were going to check out fishing holes in the little river that runs by our farm. We call it Tillasana. That's the name it came with about 10,000 years ago. A native name.

At certain times of the year, it has lots of salmon or steelhead trout returning from the ocean. They're coming back to spawn. And the fishing can be pretty exciting.

When it's low, like in summer, it's a good time to take a close look at fishing holes. For future strategies and techniques.

We decided to start with a fishing hole near the end of our farm. We followed a trail through trees I had planted about twenty years earlier. Fir and cedar. There was an understory of sword ferns and it was a favorite winter bedding area for deer and elk.

This fishing hole had a very large boulder at its upper end, maybe it was even bedrock. I'm not sure. A deep hole was just downstream from this rock mass. That was the fishing spot.

We walked through the small forest and looked into the hole. There was something in it. A body! Obviously dead. Lying on its side in the fetal position. It was a man, not old either. Only wearing socks, underwear, and a t-shirt. I didn't recognize him, but then I had no experience identifying dead bodies under water. Anyway, we were both freaked out.

We soon found one of our phones, and that wasn't easy considering we were totally flabbergasted. I called 911 and went out to meet a sheriff's deputy on the roadway. A crew showed up. Several deputies, the coroner, and a water rescue team. It was more like a retrieval team since no rescue was going to happen. Anyway, you could wade to the body.

Several of the team brought him out of the water and we all helped getting him up the steep bank. This truly was dead weight.

I didn't recognize him, either as a neighbor or friend. That was a relief, but he still belonged to somebody. Son, brother, friend, or lover.

The deputies immediately pointed out that people on certain drugs feel hot and strip down. They felt there were drugs involved. His lower legs had lots of cuts like he had crawled or stumbled on rocks in the low water. That would mean he had drowned in the fishing hole. Because there was a small tree that had fallen across the water just upstream, his body had not floated here.

He had no identification, but the coroner found a large

tattoo on his arm that was an apparent name. A last name, and one the deputies recognized. He had a history as it turns out. There were needle marks visible.

It was fascinating for Mack and me to watch the coroner examine the body and point out her observations. There were no signs of trauma. No signs of gunshot, stabbing, or blunt force either.

Finally, the coroner finished what she could do. The body was put in a body bag and hauled away to a morgue. We never heard another thing about the man or a subsequent investigation.

Needless to say, fishing on our small river has never been the same. There have certainly been a lot of fish caught. But every time we walk up to a fishing hole, we look for a body.

No more found yet, but we're still looking.

7

———————

irplane Flights I Don't Want To Take Again

ALASKA

Ed and I were flying to Alaska. Maybe he was really taking me, since he had been stationed near Fairbanks while on active duty in the Army. So, he knew the place. I certainly did not.

Alaska is this huge and wild place. Mysterious, enchanting, maybe a little scary, too. There are a lot of reasons to go there, and I was anxious to visit.

We both got time off from our jobs and met in Seattle for the flight to Fairbanks. This was a long time ago, and the main plane flying was a Boeing 707. This was especially true for anything of a distance. The 707 was a four-engine plane. It was considered the state of the art at that time and very reliable.

We loaded late in the day, found our seats and settled in.

It wasn't a full flight, so we had room. Anyway, the seat space was much larger then. It's strange, if not just outright disturbing, that seat space has shrunk, while people size has enlarged.

Ed tried to maneuver his armrest. It came off in his hand. We both chuckled. We also wondered what that could portend.

Quite a bit as it turned out.

Everybody knows, at least those who fly, that once a jet breaks loose from the ground and becomes airborne, it continues at a fairly steep and steady uphill climb until it reaches a special assigned altitude. Usually between 30,000 and 40,000 feet.

Well, this didn't happen. Well, at first it did. Then it stopped climbing. Then it started climbing again. Then it seemed to go back down. Then up. Then level. Then down.

Although hardly experts, we knew something wasn't normal. But we weren't sure what. The pilot seemed to have no intention of sharing information with us except to tell us to keep our seatbelts fastened. That wasn't very comforting either.

At some point in this meandering flight, the plane seemed to have turned. Maybe we weren't heading to Alaska anymore? Maybe we were returning to Seattle? We couldn't figure it out. Nobody was telling us either.

Since there were a lot of open seats, we decided to each sit at opposite windows. That way we could tell each other what we saw. Maybe, "There is a mountain just ahead," or "The trees are getting close." Or maybe even, "Hey, you've been the best of friends. Too bad this isn't working out."

Then, after not seeing impending mountains before us for some time or even close-up trees, lights started to appear. Lights on the ground. This had to be civilization, a city.

Then we were landing. Back in Seattle. We also had an extensive greeting party. The entire runway was lined with emergency vehicles, fire engines, ambulances, and who knows what else. It was startling. Actually, it was scary as hell. Things were worse than we had ever imagined.

But we landed safely and intact. We were herded quickly and under strict control into a side room in the airport. It's one of those rooms that you see the door to and never the inside. Once inside, the door was locked, and the alcohol was free. Imbibing was encouraged too. After all that, it did not require any encouragement, however.

Finally, we learned the plane had lost two engines. They quit working, period.

A new plane would be ready in a couple hours. Then we could load up and really fly to Alaska.

Ed and I were determined to go and got back on the new plane. Not everybody did. In fact, a lot of people just went home. They were done flying for the day, or a long time, or maybe forever.

Fortunately, the new flight was great, and we had lots of room to stretch out.

THE ARCTIC

I hadn't flown in a floatplane before, but that was going to change soon.

Four of us were on our way to Alaska and eventually the Arctic part. Me, of course, but also my brother-in-law and sister-in-law, John and Marla, and John. John was Ed's friend and came with high recommendations. Army man, Special

Forces training. He seemed a perfect companion in a wild and perhaps dangerous environment.

We figured our biggest worries would be grizzly bears at night. Well, it wasn't to be. First of all, there is no night in the Arctic in the summer. At most a quiet period. More like a time out. And John, Special Forces John, snored so loud in the tent we shared, I couldn't hear anything outside. A bear could have attacked, killed, and devoured my in-laws, and I would have never known.

Maybe the snoring kept the bears away.

We flew from Seattle to Fairbanks on a big jet and stayed overnight in the airport. The next morning we flew to Fort Yukon. This was a much smaller plane. Old, too. A DC-3 left over from World War II. These planes flew for years after the war. It was filled with locals, their new purchases, new dogs and whatever. Across from my aisle seat, a native had his small new refrigerator in the seat next to him, and it was secured with the seat belt. Puppies could be loose and friendly. Older dogs, wary. As were we.

From a lake just outside Fort Yukon, our floatplane awaited us. The pilot was a Vietnam vet. We figured if he had survived that, we would be fine. He could only fly two at a time, plus gear. We loaded up. Future tent mate John and I went first. I waited for the pilot to go through a checkout procedure. But no, he just started up the engine and began moving across the lake. "I guess he knows," I thought.

About halfway down the lake, the plane tipped to the side, or maybe he tipped it. I wasn't sure, but I was pretty sure we were going to tip all the way over, crash, and drown in the lake. Just before that happened, he tipped the plane the other direction. He was breaking the pontoons free from the water one at a time. Apparently, the small plane didn't have enough power to break both pontoons free at once.

Once free, we skimmed just above the trees at the end of the lake and were on our way across the Yukon Flats. The Flats are enormous. Filled with lakes, rivers, tundra, some trees and not much else.

We were flying over them on our way to the Brooks Range and the Arctic National Wildlife Refuge.

It wasn't long before something started coming up from the engine. A liquid, it seemed. Maybe oil. It wasn't long before it pretty much covered the windshield. Then the pilot opened his side window and tried flying looking out it. Naturally, that didn't work very well. He couldn't make out much in front of the plane anymore.

John and I were more and more disturbed. "What the hell was going on, we wondered?"

Suddenly the pilot said, "I've got to put her down." Meaning the plane. And us, of course. Since we were in a lake-littered landscape, it would be easy to find a place to land. He spiraled the plane down and suddenly we were on a large lake. There was a hut at one end with smoke coming out the chimney. A person emerged and stared at us like an alien ship had landed.

I was thinking that if we had to spend the winter, maybe we could stay with him.

Before we knew it, the pilot was out, opened the engine cover, wiped a few places with a rag he had, and was back in the plane. He started the engine and proceeded to take off.

I looked at him questioningly. He replied casually, "I had spilled some oil earlier. It's ok now."

And then we were in the air to our destination.

Mongolia

This is a lonesome land squeezed between Russia and

China. It's also large, has few people, and really only has one city. Ulaanbaatar. There are a few scattered villages, maybe even small towns. Roads are rare. To travel to distant towns, flying makes the most sense. Driving was too unpredictable, especially if you couldn't find a road.

Bill and I were flying to a western town with a name neither of us could pronounce. Our goal was the Gobi Desert and wild Bactrian camels. The first goal we achieved. We never found the camels. Only tracks and poop.

Now we were sitting in a packed WW2 Russian aircraft. It was waiting in line to get started by an electric generator several planes ahead of us. Apparently, it had no startup abilities of its own.

There was no airflow. There was no cooling either. And this was taking a long time. An hour, maybe a lot more. We all started sweating. A lot. Profusely. The air was full of sweat and water. I'll skip the odor part. The moisture was moving to the ceiling. There it formed condensation.

Finally, our plane made it to the jump-start station. The engines started and so did the cooling mechanism. We were almost suddenly comfortable, but our clothes and the air were water-filled.

Soon the plane gained a lot of altitude. We were off to a small town at the edge of the Gobi Desert. The one with a name we couldn't pronounce.

As we gained altitude, the cabin cooled, a lot, as it turned out. The moisture froze to the ceiling. It turned white like it had snowed, but upside down.

At least it was out of the way, and we were certainly more comfortable. We thought so anyway. We were pretty much dried out.

It went on this way for some time. Finally, we neared the small town, and the airplane began its descent. As we got lower, closer to the ground, the air warmed up. The upside-down snow began to thaw. And there was a lot of it. It thawed and turned to rain. We were now in a rainstorm.

By the time we landed, everybody was soaked. Wetter than ever. But we were at the edge of the desert, and the warm afternoon sun dried us out fairly soon.

We hired a Mongolian with a Russian jeep to drive us back, however.

SOUTHERN CALIFORNIA

The other stories might have scary parts, but they also had dark humor. They only involved adults. This one was really scary and involved my kids traveling with me. There was no humor, dark or otherwise, involved.

My father lived in Southern California, which was a long way to drive from our home. Too long. So, for visits I would fly down. It just took a couple of hours or so depending on stops.

That's how I took our kids too. My son and daughter. They loved it. They got to visit their grandfather and, almost just as important, visit Disneyland. And the weather always seemed good, especially if you lived in the Northwest. So, it was always a win-win-win situation.

We had just finished a delightful visit for the kids and the adults. The flight included a quick stop at another city before we flew north. We were sitting on the right side of the plane in the first row. My son was at the window. He wanted to see things outside. I sat in the middle, and my daughter sat by the aisle.

That way I could see them easily and maybe prevent potential brother-sister conflicts.

It was working just fine, and we were starting to descend

to the next stop. We were perhaps a thousand feet or more above the ground.

Suddenly, our plane dropped sideways to the left. It was instantaneous. There was no time to think. There wasn't even time for your life to go before your eyes. There was no time to make a sound, even a scream.

Just as sudden, the plane righted and continued on to the landing. All the passengers were, let's face it, freaked out. Not a word was said by the pilot or steward or stewardesses. Not a word.

I didn't know what to think, except we weren't going to die now. My son, sitting at the window, turned and said, "We almost hit another plane."

We landed; no more excitement either. We waited until the plane was empty. "My son says we almost hit another," I said to the steward. It was really a question. He said it was true and tragically too common and mostly caused by private pilots flying in the wrong areas.

I never flew our kids to Southern California again. I couldn't bear to tell my father what happened. Instead, I encouraged him to visit us, which he did.

8

———

C upcake

THIS IS NOT about a dessert item. This is about a pony, a Shetland pony. Her name was Cupcake, and she was supposed to be the perfect little girl's horse. For our daughter Alissa. Well, that didn't really work out.

But first, about Shetland ponies.

Years ago, in fact, many, many years ago, Shetland ponies were very common. It seemed as though everybody was raising them. They had become one of many animal pyramid schemes. They were one of the first.

An animal pyramid scheme works by raising and selling animals that will reproduce and then be sold to another person who will do the same. It always requires more animals and willing buyers. Some people actually make some good money, most don't.

There's a long list of these animals now: Shetland ponies, chinchillas, Arabian horses, ostriches, llamas, and alpacas. Different dog breeds. There will be more, too.

Eventually, the scheme loses momentum and crashes. Maybe that happens when enough folks figure out what's happening and decide not to buy and perhaps lose their money.

A Shetland pony is as cute as a bug, as they say in the country. Beautiful flaxen manes and tails. Small, easy to have around, handle and feed.

After the scheme tumbled, they were everywhere. Cheap too. I bought one for $25 and got the other free. Cupcake and Dolly.

I bought Cupcake shortly before Christmas long ago. I kept her at a neighbor's barn until Christmas Eve. Then I snuck her home. I also had bought the cutest little saddle you could imagine, along with a miniature saddle blanket and bridle and bit. They all got packaged and wrapped together for Christmas morning. It was a big box.

On Christmas morning, Alissa opened the enormous box. "Wow, I got a saddle!" she exclaimed.

"What goes with a saddle?" I asked.

She hesitated.

"Dig deeper," I suggested.

Out came the little saddle blanket.

"Dig even deeper," I suggested again.

Out came the bridle and bit.

"What's next?" I asked, getting impatient. But by then she was off to her other presents. I settled for a cup of coffee and waited.

It took a while, but finally all the presents were opened, played with and put aside. Then she crawled in the big box. Perfect little house for make-believe.

Sometimes I wonder if kids enjoy the boxes as much as the presents that were inside them.

Finally, she seemed done, and it was time for me to feed animals in the barn. I asked if she wanted to come help me. She did.

There she found Cupcake, waiting. Needless to say, Alissa was really excited and surprised that a pony went with all those gifts. Who would have guessed? That, unfortunately, was the high point of her pony experience.

Later that day we did saddle up Cupcake. I led the two around our place. That went ok. It was later when things fell apart.

Finally, we decided we could turn the pony loose with my daughter riding. Under close supervision, of course. That went well for a few minutes.

It soon became apparent that Cupcake was quite experienced. Not at being a riding horse, but at removing riders. Since bucking takes work, she used swiping instead. Usually this involved walking up to a low limb, walking under it, and swiping off the rider. She was so quick and efficient I realized this was learned long ago.

Later, we acquired another pony, Dolly. Same talents. Obviously well experienced, too.

Pretty quickly, Alissa lost all interest in horses. For a few years, as a matter of fact.

I did find a home for both ponies, with George.

George had grown up working at cattle ranches in the rugged and dry Tehachapi Mountains in Southern California. He had moved north and found a mill job. But the cowboy had never left him. He still had horses and cattle.

George also had another passion. Rodeos. He decided to set up a rodeo for kids. He needed some animals. It was a match made in heaven. I needed to get rid of two demonic ponies. He needed bucking stock.

Cupcake and Dolly had a fine career. There was a little more work involved, but it included room, board, and medical care. It suited their personalities just fine.

9

M ore About George

THIS IS ABOUT GEORGE, the cowboy from the Tehachapi Mountains.

As I mentioned before, at some point, George moved north, bought a small ranch along the Columbia River, and got a steady job at the mill. He raised some cattle and horses. There was no way he could get that out of his blood. That's how I got to know him. Caring for his animals.

George was short and stocky, really stocky. Not only did he cowboy and break horses, he rodeoed too. He had the kind of body that could take a beating. For a number of years anyway. That's probably why he decided on a mill job. Nobody's body holds up forever.

He never lost his interest in rodeos. That's why he decided to set up a kids' rodeo circuit.

And that's how I got rid of the two demonic ponies I had given our very young daughter.

As it turns out, there were a lot of those ponies out there to be given away. Pretty soon he had quite a collection. He was going to have a big bucking string for the kids.

Among the new group of ponies were stallions that needed to be gelded, or neutered. After he was done collecting, George called and made an appointment for the surgeries.

There was something about George that I didn't know, but we'll get to that in a moment.

We had several little stallions to geld. It was going well. Two down, two to go. George was a great hand. That was no surprise.

We had the third pony down, anesthetized. George was holding the leg rope. I had just started my surgery. For some reason I looked at George. Suddenly his eyes rolled back in his head, and he fell on the pony. He was fast asleep.

It turns out George had a form of narcolepsy.

I was stuck without him. I started yelling and yelling. Finally, George woke up. I guess he heard me. He looked at me with a sheepish grin and went back to work. Just like nothing had happened. I guess he was used to it, so he just carried on.

We finished that pony and others without any more sleeping spells.

10

F lying With Joel

HE LIVED across the big river, the Columbia, a mile or so north of it. I lived on the other side, the south side, about a mile or so away, too. This is a really big river, maybe 2 miles across. We weren't really far apart as the crow flies, as they say. Or, as it turns out, by a small airplane either.

Joel had a small dairy in a river valley. Not only was it beautiful, tranquil, and definitely off the beaten path, it even had a covered bridge, still functioning, left over from years ago.

I would make occasional trips to Joel's farm to treat sick cows. Sick cows are sooner or later inevitable. I enjoyed Joel and caring for his cows, but it was a long trip. Really long trip. I needed to drive downriver to the next town, cross a bridge several miles long and then wander up a small highway to his place. It took time, a lot of time. Sometimes I was just too busy to make the trip. This became a problem for both of us as time went on.

I'm not sure when it happened, but Joel started flying a small airplane. Maybe he just went back to something he had done years earlier. Or maybe it was from one of those dim dreams we all carry around.

Anyway, it wasn't long before he got a license, which meant he was pretty good at it. Or I hoped so.

And then it wasn't long before he bought a small, single-engine plane. Pretty small really. Of course, he told me all about it. Joel thought the next time he needed me he would just fly over and get me. I was intrigued, but not real sure about the idea.

We didn't have an airport nearby, but we did have an airstrip. It was marginally kept up, with broken pavement, weeds growing through, and a herd of grazing elk to chase off. An incoming plane would need to buzz the strip several times to chase off the elk. In spite of that, it worked pretty well.

Only one pilot had crashed in recent years. He failed to clear the trees taking off to the west. Unfortunately, wind conditions almost always meant you had to take off to the west. Sadly, it was his last flight. But since he was really older, people said he shouldn't have been flying anyway. So, it wasn't the airstrip's fault, they said, and we all believed it.

Finally, it happened, Joel had a sick cow. He was excited to fly over to get me. I was less excited but willing to give it a try.

We met at the airstrip a few mornings later. I then realized his plane was really small. I squeezed in and held all my tools and supplies on my lap. I had to leave a few things behind.

The takeoff was fine. The trip across the river was quick,

and we were soon at his place. Well, actually the neighbors. His land was too bumpy to land, he told me. So we landed next door in a cow pasture along a barbed wire fence. It was a bumpy landing, the plane bounced all over the ground until we stopped. I thought we might end up in the fence, tear a wing off, or who knows what. It made me wonder what his land was really like.

For Joel, it was routine and not a problem. For me, I could only think about what the takeoff was going to be like. If we would even get off the ground.

The cow was waiting for us, locked up in a stanchion. It didn't take long to examine and care for her. Anyway, I was anxious to get home. Anxious to get back in the airplane and try to take off. Anxious for me to meet my fate, just get it over with.

As it turns out, everything went fine. I begged off from future flights, saying I could not carry enough of my tools and supplies to do my job.

11

G etting Lost In Your Own Backyard

ACTUALLY, my backyard is a small farm. Twelve acres. So not a normal-sized backyard. It has a house, barns, lots of trees, and pastures. A small river on one side and a gully on the opposite side. The gully becomes a small creek in the winter. Then, we call it Winter Creek. It is, quite frankly, a jungle of brush, berry vines, and assorted trees.

Even though it was 12 acres, I knew it well. Or I thought so. Anyway, I had lived and wandered through it for 40 years. You would have thought I would know my way around.

Abby was my dog then. A black Lab. Over the years she had become quite territorial. Especially around the house and barns. Largely, that meant she didn't like raccoons on her turf. I agreed with her. At that time, we had free-ranging chickens, which too often ranged into the digestive system of raccoons.

She would find raccoons at night when out patrolling. Then she would tree them and start barking. It was a special bark. That was my call to come finish the process. It was a pretty good system for years.

Finally, however, I found a better system. Keep the chickens inside.

It was almost winter; the jungle had not transformed into a waterway yet. Of course, it was dark. Abby started barking somewhere in that jungle. I knew she had treed a raccoon. I also knew she wouldn't leave until I arrived and finished the process.

Finding her and getting to her was a job of listening to try to locate her and beating the vegetation so I could get to her. It took some time wandering around.

Finally, I found her and her treed enemy. I finished the process.

We were ready to return. Abby started off in one direction. I called her back. She was going the wrong way. That didn't make any sense to me. I always thought dogs were pretty foolproof with their directions. After some arguments and me getting very bossy, she agreed to follow me.

It took a lot of brush beating. A lot more than I thought it should. And a lot more than it took to find her and the raccoon, initially. Finally, I pushed through the last of the thick brush and was in the open.

I quickly realized two things.

I went the wrong direction and ended up on the neighbor's property.

And the dog was right the whole time. She knew the way home.

Now I had to beat my way back through this jungle.

This time I let Abby be the guide. Except for the extra effort, it worked out much better.

12

———————

n Arresting Development

THIS WAS GOING to be really adventurous and exciting. My two brothers, Rick and John, and my young son, Matthew, and I were traveling to Alaska. To the Arctic. To kayak a remote, wild river in the Brooks Range. That's a large mountain range that circles across northern Alaska before it disappears into the Arctic Ocean. I had been there before with my son. But this was going to be a family trip. My brothers with me and my son.

We took a lot of gear. Two big kayaks that folded into a couple of gigantic suitcases each. A stove. Food. Tents, sleeping bags, pads, clothes for all situations. Fishing gear. And importantly, a rifle. Grizzly bear protection. It was a .45-70 lever action designed for killing buffalo, originally. I figured it would work on a bear. I also very much hoped I never needed to test that idea either.

In those days traveling, especially with excess baggage, wasn't such a big deal. They just loaded your stuff and off you went. They didn't even make a big deal about guns and ammo.

The evening flight from Seattle was loaded with oil pipeline workers. The Alaska oil pipeline was under construction then. Mildly put, they were a rowdy bunch. The truth is, well, unprintable. They largely stayed in the back of the plane, raised hell and harassed the stewardesses. It seemed that there were no stewards in those days. That's probably just as well, as they probably would have ended in fights and most certainly not gone well.

It seemed the stewardesses had a plan. Free drinks, lots of them and fast. It was tricky, of course. At a certain level of drunkenness, these guys were worse than ever. So, the women had to get them quickly into a semi-somnolent state. Especially if we were going to make it to Fairbanks without a riot.

These women were experienced and their plan worked. We arrived safely without a fight, riot or structural damage.

Regardless of their success, it was the most entertaining flight I've ever taken. Certainly better than many of the movies I've watched over the years.

It was late at night when we arrived in Fairbanks. Not really dark, but dim. After all, this was the Arctic. And it was midsummer. The airport was pretty quiet. We found an out-of-the way spot and put our gear in a pile. It was more like a small mountain actually. We planned on sleeping, or napping with our gear until time to catch our early morning flight to Bettles.

We had no sooner settled down than a man approached us. He had a baggage storage room next to where we had piled our gear. He also offered to store our small mountain of gear for a reasonable price. It sounded too good to be true. And, actually, it was, as we learned later.

Our big, really big, concern was when the storage area would reopen. We had a very early flight in the morning. He assured us several times that would not be a problem.

We finally believed him and moved all our gear into his storage area. Now we didn't have to worry about it and could maybe nap some. That, of course, seems impossible in an airport based on my experience. But at least we didn't have to worry about our equipment.

After several non-sleeping hours, we got up to retrieve our gear. There was already a line. Maybe 20 more people with early flights.

Nobody showed up to open the storage area. We waited.

Nobody showed up.

Still nobody showed up. The time was nearing for our flight departure. Others' too.

Finally we made contact with an airport official. They called the owner. No answer. They tried some other possibilities. No luck.

Everyone was getting concerned, anxious for sure. Maybe some irate, I'm not sure.

I noticed my brother Rick looking over the door, which was, of course, locked. But he was looking at the metal grate covering where a window would be in a normal door. It was held in place by screws with Phillip's heads. Perfect for our Swiss Army knives, it seemed. We went to work. Nobody discouraged us. In fact, they were looking forward to getting in and getting their luggage. Actually, they were encouraging us. By then, everybody waiting was irate about the situation.

It didn't take long, the grate was off. I reached through the opening, unlocked the door and opened it. Everybody rushed in and got their luggage. Then they were gone. After all, they had flights to catch and not much gear. We had a flight to catch, but we also had an enormous amount of gear. It was going to take us a lot longer.

When I opened the door, I triggered an alarm, a burglary alarm. And that brought the security force. Of course, all the other passengers were long gone. Brother John and son Matthew were hauling part of our gear to our next boarding area. Hopefully, our next boarding area. Rick and I were busy bringing our remaining gear out.

Then the police arrived. I explained our dilemma. They were unsympathetic. Then they explained our new dilemma. I was under arrest. Actually, I think they used the word "detained," which seemed a bit more polite. They also explained this was a class C felony.

At least I wasn't cuffed.

But they needed the owner to press charges, so an opening did exist for me.

I got my brothers out of the area. Rick was known to possibly have seizures if put under high stress. I was pretty sure this would qualify. John had an unfortunate previous legal encounter and didn't need another one.

I wanted my young son with me. I was thinking up a plan to get out of all this. I explained a few things to him. He was game. He was especially game after I explained he might end up in a foster home somewhere in Alaska. Nothing like putting a little pressure on for some unre-hearsed acting coming up.

Finally, the owner showed up. A rather disheveled middle age woman. But I noticed kind, if not sympathetic, eyes.

The police explained what happened, my crime. Matthew started crying, right on cue.

They told her she had to press charges since it was her place I had broken into.

Matthew was crying even more.

She wavered. He kept crying. At that point I didn't do much. I was going to let it play out. Which it did.

The inner mother won out over my crime. She said she would not press charges and we were free to go.

Then, at the very last moment, she apologized for her failure to have her storage area opened as promised.

Obviously, we accepted the apology, moved our remaining gear, got a new flight and left as soon as possible.

In fact, we left absolutely as soon as possible just in case the owner came to her senses and changed her mind. As it turns out, she didn't.

13

F

alling Out The Front Door

I WAS BETWEEN JOBS. I had just stepped away from a busy vet practice. I was even a partner there. It was not only interesting and challenging, but busy. Too busy. Its rhythms were wrong for me. I wanted to do more than just work all the time, day and night. I needed to sort out what I wanted and how to do it.

Life was cheap then. Really cheap. We bought a small house on a few acres near the Eel River in the redwoods. There weren't any vets anywhere near, so I started up a traveling clinic. We could live on 200-300 dollars a month. Ridiculous by today's standards.

It was also interesting times.

West of the river, communes were developing. Young folks from the LA and San Francisco areas were moving north. A very idealistic group, often well-educated, creative and surprisingly hard-working. They were buying land, building houses or cabins, starting gardens, and raising animals.

Raising animals is where I came in. I would visit the little towns, villages and hamlets weekly. They knew when I was coming. Cattle, horses, sheep, goats, pigs, dogs, cats and poultry. I took care of everything. Many of them had little knowledge of farm animals, so I spent a lot of time teaching basic care of their animals. I set up and taught some classes too.

To the east of the river were traditional ranches, cattle and sheep. Naturally, they also had horses. The ranches were large and long established. I was used to this. These people had been around and were very knowledgeable.

It all made for an interesting lifestyle for a few years.

On the west side, the commune side, marijuana cultivation and use were pervasive. Most of it was pretty harmless and largely for personal use. The varieties were quite mild by today's standards. But the newer varieties were to arrive soon.

Occasionally, I would be vaccinating a dog that was hitchhiking across the country with its owners. And they were taking a hundred pounds or so of dried marijuana to sell. That sounded pretty dubious to me, but they usually made it.

In later years, the marijuana cartels moved in and everything changed. It got ugly and just plain dangerous too. We were long gone by then.

One day, Jeremy stopped by one of my country clinics in a local hamlet. He dropped a joint of marijuana in one of the drawers of my vet box. The drawer with eye and ear meds. I still remember.

"It's the new stuff, from Thailand, Thai dope, crazy

stuff," he casually mentioned. I didn't pay much attention at the time.

Some months later, in the summer, we were sprawled out on our large bean bag chair. It was after dark and still very warm out. The kids were asleep in their beds. It was very relaxing. We didn't have a care in the world. We had the front door open, wide open.

The stereo was playing, I'm guessing the Jefferson Airplane. Grace Slick was singing.

Eventually I remembered the joint of the new marijuana. I grabbed a flashlight and went looking for it. Easy to find, as it turned out.

"This is supposed to be pretty crazy stuff," I said as I lit up. We smoked about 1/3 of it each. Then the record ended and I got up to put a new one on.

"I don't really feel anything different," I said as I went to the stereo. By the time I got there, it had hit me full blast. I couldn't figure out how to change the record. My wife felt it too and called me back. We sat together trying to figure things out, which we couldn't do.

About the same time, we noticed the open front door. Suddenly, both of us thought we would fall out the front door. Yes, fall out the front door. I decided to close it, to protect us. But I was afraid to stand and walk, well, because I might fall out. So, I cautiously crawled to the door and closed it.

Now we were safe.

For a long time after, I stuck to drinking beer.

One Country Song Each

Warning. This contains some graphic details. However, if you're used to farm and country things, it's almost normal.

George needed his stallion neutered. George certainly did not need a stallion. Most people don't. It just took him several years to figure it out. Several years too many. That means the horse wasn't so small anymore. A little more complicated.

In case you don't know what neutering means, it means castrating. Or with horses, it is often called "to geld." To create a gelding. A more sensible and gentler creature. Here's the graphic part, or one, anyway. It means surgically removing the testicles.

It's accomplished with a general anesthesia given as an injection. Almost always given IV. It acts quickly and is effective long enough to produce a peaceful moment for the surgery.

Once a testicle is exposed, a special tool called an emasculator is used. It has two adjacent parts. Cutting and crushing. The crushing part stops the bleeding. The cutting part, well, it's obvious. It removes the testicle.

George lived back in the hills. I'm not sure I had been there before. It took a while to get there and find it. He lived on a hill. There were no level spots, so this would be done on a slope. Not ideal, but doable.

We found the mildest slope. Hopefully the horse wouldn't roll away during the surgery. It didn't.

I had Sarah along, the new vet who had just started working with me. We were still working together. I would turn her loose before long, however. Gelding horses, especially older ones in questionable situations, requires some mentoring. And that's why we were together that day.

We had everything together. The drugs. The big rope to pull a leg back once the horse was asleep. The surgical equipment. We were ready.

Still learning and, of course, curious, Sarah had a question. She wanted to know how long to apply the emasculators. And that's important to control bleeding.

Actually, it's sort of subjective for me. Less time for smaller horses. More time for larger and older horses. I think I learned on the run over the years.

So I asked what she just learned at school.

"Keep it on for a country song. Each," she answered.

That perked George right up. "I'm a country singer," he announced.

It turns out he sang at local bars around the area and had been doing that for a long time. Before we knew it, and before we got started, he was off to his house close by. Just as quick, he returned with his guitar.

"I'm ready when you are," he said.

We proceeded with the surgery. George proceeded to sing at the appropriate time.

The surgery was successful. We got some country music with it too.

Pretty good music, I might say.

15

G ypo Logging

RALPH HAD a small herd of cattle. That's how I got to know him. Taking care of his cattle. But he had another business too. He was a logger.

Ralph was a one-man show. He had a one-man logging operation. That meant he did it all. Fall the trees. Cut them up. Load them. Take them to the sawmill. And then clean up the logged area afterwards.

It's a dangerous business. In fact, logging is one of the very most dangerous of all occupations.

Guys like Ralph were called gypo loggers. A weird term, for sure. Originally it was derogatory. It meant they were gyping the union loggers by working around union contracts. They were not in unions.

In recent years, it changed. It became a badge of honor, being a gypo logger. It meant you were a strong, independent, hardworking, rugged individual.

Ralph finally got to the point where he hired fallers to drop the trees. He was a bit older. Anyway, that was usually the most dangerous part.

He still did all the other things, however.

Ralph had just finished a small logging operation south of town in the hills nearby. The logs had been taken to the sawmill. He only had the cleanup left.

Clean up meant using his bulldozer to push the trimmings into large piles for burning. These were the leftover limbs, and sometimes they could be quite large. Usually, the burning took place in wet months to lessen fire danger.

Once the cleanup happened, the forest could be replanted with small trees. Pretty routine.

This time something went wrong. While bulldozing the branches, one snapped back. It hit his head. Even worse, it dislodged an eye. Somewhere in the tangle of limbs was his eye. Hopelessly lost. Hopelessly damaged too.

He was bleeding. A lot. Ralph used a rag wedged under his seat to stem the blood flow. He pressed it against his empty eye socket.

He got off the bulldozer. Briefly looked around for the missing eye. Without luck. He got to his pickup and started for town and the hospital.

He started driving. Things were going as well as he could hope for. Not too much bleeding. But suddenly, he got hungry. It had been a long time since he had eaten, and this accident was taking a lot out of him.

Ralph stopped at a country store nearby. It's one of those places that sells everything. Groceries, gasoline, drinks, sandwiches, some hardware, fishing and hunting gear, and who knows what else.

He bought a Coke and sandwich and consumed them quickly in the parking lot. Then he was off to the hospital.

The store owners were probably used to maimed loggers over the years and didn't think too much about him holding a rag over his eye.

Ralph made it safely to the hospital. They stopped the bleeding and stitched his eye shut. He was good to go, for a logger anyway.

That probably wouldn't have been true for the rest of us.

16

———

He'll Be Back

IN YEARS PAST, a lot of rural counties not only had jails but farms for the prisoners. In some cases, the farms were large and similar to big commercial farms or ranches. But more often, they were small. Maybe not more than 20 or so prisoners. With an assortment of farm animals. Some cows, some sheep. Not usually a dairy. That's too complicated. Pigs were always a good bet. A dozen sows having piglets twice a year would keep the prisoners pretty busy too. And produce food for all the inmates. That seemed like a win-win situation.

Times have changed and so have the prisoners.

In those days there were no hard drugs to hopelessly disable the inmates. And there weren't many guns either. Very few actually. The violence level was much different.

These prisoners weren't really evil people. Some would

barely be considered criminals by today's standards. They were mostly screw-ups. Probably got drunk and did something just plain stupid. And some just kept doing it, too.

My boss thought I should be the vet for our local prison farm. He wanted to care for the expensive horses and the big cattle ranches. Totally understandable.

I'll admit that I wasn't really sure what to expect. I had no experience with criminals or convicts. I wasn't even sure if it was safe. I was a bit apprehensive as a young guy mixing with older criminals.

But it turned out ok. After my first visit, the prison officials just turned me loose with the inmates. They were just fine, barely criminals and certainly not dangerous or evil. In fact, I looked forward to my visits, which was mainly for the pigs. These prisoners were surprisingly congenial and, quite frankly, fun to be with.

There was Bobby, Lenny and Mel. They were the regulars. Others, too. But they came and went. These guys knew their pigs. They enjoyed their pigs. They took really good care of them. If you've spent much time around pigs you will understand how they felt. Pigs are smart, really smart, and they have personalities. Plus, they can actually be sort of lovable.

It hardly seemed like a prison to me. And maybe it really wasn't when compared to being in a cell. These guys got to be outside all day, or at least in a barn. No comparison for sure.

And just maybe, this was a pretty good life for them. Especially since some seemed prone to mischief outside of jail. Things were stable and consistent here.

I'm sure they missed a few things though. Like alcohol, which usually got them imprisoned. Or maybe women. Or maybe families.

One day I was visiting the prison farm and working with Bobby and Lenny. We were castrating baby pigs. One caught, one held, and I did the job. It gave us lots of talking time. But I noticed Mel wasn't helping or even around. I thought that a bit strange. I had gotten used to the three of them.

"Where's Mel?' I asked.

"He got out," Bobby answered.

"But he'll be back soon," he followed up quickly. They both laughed and laughed.

17

———————

Hiding In Plain Sight

IT WAS a great place to keep your horse. A large, indoor, all-weather riding arena.

Large pastures. Nice and comfortable stalls. And outdoor places to ride, too.

There were 20-30 horses. Mostly owned by women, some were mother and daughter combinations. Katy was the exception. No mother, just alone.

I was there often. With that many horses there was always something to do. Vaccinate a new horse, look at a bad eye, suture a wound, examine a lameness. Assorted sick horses too. Maybe even check out a newborn foal.

Almost all the horses were purebreds of different breeds. Arabs, Morgans, Paints, Quarter Horses. A few Thorough-breds. Katy's was, well, some mixture. She, or her family, didn't have the resources to buy a fancier horse.

Since I often saw her there during the day, school hours, I wondered if she even attended.

I thought more and more about a friend and rancher who had told me about his chaotic childhood. "Nobody raised us. We just grew up."

"That" is going to create a lot of independence or a lot of disasters, I thought.

It did just that in his family.

You would call her tough. Maybe rough too. Short, stocky, lots of muscle. A bit chubby too. Sometimes more than others. Good with her horse too. Really good.

I hadn't seen Katy for some time, maybe several months. I asked one of the horse women about her.

"You haven't heard?", she replied.

"No," I answered.

"Let me tell you. It's quite the story."

As it turned out, Katy started to seem a bit chubbier to everybody. But nobody thought too much about it. She had that kind of build and her weight varied anyway.

One night Katy started having abdominal pains. The pains kept getting worse. And worse. And worse.

Finally, she told her parents she was having terrible abdominal pains. They immediately took her to the emergency room at our local hospital. The doctor on duty examined Katy and proclaimed her not only pregnant, but also in the process of childbirth. She was having a kid.

She had a healthy, normal boy. And now she was a mother and a horse owner.

Naturally, everybody wondered how that could happen. How could you hide a pregnancy from everybody? Maybe even hide it from yourself.

Nobody asked, however. She now had plenty to do without answering our nosey questions.

18

C ollecting Car Insurance

AFTER YEARS of buying and driving used cars, it finally happened. We bought a new car. Not just new to us, but new straight from a car dealer. Brand new, unused. It was pretty exciting.

No more worry about parts failing. At least not for a long time.

Good tires too. No more cheap retreads either. Retreads that could unravel at any inconvenient moment and leave you stranded.

We were in the process of moving from the big valley in California to the very north coast. We had an empty weekend, so we took a drive in our new car into the Sierra foothill country.

It was a hot afternoon, and we decided to stop at a grocery store for drinks. I think it was a Safeway. Because we were in foothill country, nothing was flat. The parking lot was on a hill.

After getting back to our car and enjoying our drinks, we heard a scream. And a rumbling noise coming towards us.

We looked up just as a shopping cart rammed into our passenger door. It left a serious dent.

A woman had finished shopping and let her shopping cart go. It started moving downhill towards us, gaining speed until our new car stopped it. Our brand-new car, is now dented.

She got in her car and quickly drove away. We were on our own to deal with it.

But we were leaving to move north. That was just days away too. There was too much else to do. Anyway, it was just a dent. But on our very new car.

We moved, got settled in a house we had bought, and started our new life. A new life keeps you busy. We pretty much forgot about the dent.

Several months later, maybe even a half a year or more, the car got another dent. This time on the opposite door. It happened while backing out of our new garage. The new garage, which wasn't new at all, had a pole in a difficult place to maneuver past. It got hit and thus the dent.

Two dents was too much to ignore.

I contacted a new insurance agent. She worked for the same company that had covered our car before. I told her the car was dented from backing out of our new garage. We made an appointment in the near future.

I showed up and explained to her what happened. Of course, like any competent insurance agent, she looked over the entire car. And, of course, discovered the other dent on the opposite door.

I had forgotten about that when I told my story. Well, it

wasn't a story at that point. I had told exactly what happened. Now I had a problem. Two dents on opposite sides of the car. That didn't mesh very well with my story.

"Why is there a dent over here?" she asked from the passenger side.

I was trapped. For a moment. But then I replied, "You don't know my wife".

She looked at me, cocked her head, and accepted my story. Maybe she didn't believe it. She probably heard all kinds of stories. Anyway, she let it go.

We got the car fixed and were very careful backing out of our unusual garage from then on, or just left the car outside.

19

I'm Not Doing This Again

WE'VE ALWAYS HAD CHICKENS. Sometimes just a few, sometimes a lot. Usually a rooster, too. Most roosters take care of the hens. For instance, if they find something delectable, they call the hens to enjoy it. Or if there's an eagle above, they warn the hens who are busy eating bugs. Obviously, there's trade-off in the form of a little avian love life.

Most roosters are usually fine to be around. Sometimes, one will take its job too seriously and start attacking people. Little kids get the worst of this. Obviously, those roosters have to go.

One year after removing a mean rooster, the flock was rooster-less for some time. I had to find another rooster with a proper personality. Unfortunately, two alpha hens rose to the top of the famous chicken pecking order.

They could not compromise, so one moved away and took half the flock along. Unfortunately, this hen didn't think things out well and moved down by the creek. Maybe chickens never think things out very well. It was a perfect highway for things that like to kill and eat chickens. Mainly raccoons. After a couple nights of almost massacre-like killings, the few remaining chickens moved home, where it was safer.

For years our chickens ran wild during the day and would seek shelter in the rafters in the sheep shed at night. It wasn't perfectly safe, but usually so.

Chickens make tempting meals for lots of animals. Raccoons, possums, skunks, coyotes, eagles, even owls. Every few years the chickens would get raided. Usually raccoons. Usually more than one.

Then we started having more raccoons around. And they were killing chickens, more and more.

Fortunately, my dog was good at finding and chasing them up a tree. Treeing them, as it's called. Of course, it was always at night. But the type of bark she used alerted me to her raccoon treeing.

I would get up, get dressed, get my shotgun and flashlight. And then figure out where the barking was coming from.

It's pretty easy to find a raccoon in a tree. They look at the flashlight and their eyes reflect back. The problem is figuring out how to use the flashlight and shotgun together.

I was pretty sure I had it figured out. Hold the flashlight in my mouth and then aim along the barrel of the gun. Everything lines up perfectly. The light and the barrel. And everything stayed perfect until I pulled the trigger. The shock of the blast drove the flashlight back into my mouth. I was sure I had teeth knocked loose or maybe out. I was woozy from the concussion in my mouth. Finally, I got the flashlight out and checked my teeth. They were ok, not even loose. Just some blood.

The shot was perfect, though. The chicken-killing raccoon was now dead.

However, I would have to find a different way of using the shotgun and flashlight in the future.

20

J eff Stories

THESE ARE dog stories about a dog named Jeff.

Except for a brief period when I was very young, Jeff was my first dog, at least one raised from a puppy to adulthood.

He was a black Lab. A hunting dog, and that's what I hoped to use him for. Hunting pheasants, quail, ducks and geese. That didn't all work out too well, but he became a wonderful companion for years. And as it turns out, good for lots of stories.

Jeff was strong, brave and fearless. Headstrong too. Or maybe just plain stubborn. He probably would have been a good Marine if human. Jeff had a good nose for bird hunting, retrieved well, and was unbothered by gunfire. He was "birdy", as they say.

He also had another hunting proclivity. He loved to

hunt, find and chase jackrabbits. I was never able to break him of that desire.

Once out hunting quail, he jumped a jackrabbit and was gone, lost in the chase. I whistled, yelled, shouted and screamed. It didn't work. He simply disappeared after the rabbit.

I continued my hunt alone. Without success, by the way.

Sometime later, maybe nearly an hour, I spotted something black heading my way. It was Jeff, and he was hunting back in my direction. As he got closer, I realized there were three guys hunting behind him. They weren't hunting quail, either. They were hunting jackrabbits and had found the perfect dog to help them.

When he spotted me, he ran right up to me like we were the best of buddies, which was pretty much true. Somehow, he also conveyed that he had just had the time of his life.

The three hunters thanked me for letting them use my dog and I went home. With Jeff and no quail.

We didn't have shock collars in those days, but somebody suggested tying a dead rabbit around his neck until he became disgusted with it. Hopefully, that would discourage him from hunting rabbits.

It was easy to find dead jackrabbits in those days. They were commonly found along country roads. Dead. Hit by cars or pickups.

I quickly found one, pretty fresh too, brought it home and tied it around his neck.

Jeff just loved that dead rabbit. It became his new best friend. Finally, I was the one that became disgusted with it. So, I removed it and buried it.

That did not change his rabbit chasing at all.

Let's face it, Jeff was a scalawag. Here are some other things he did...

Wandered away one day and ended up on the third floor of the women's dorm on the university campus.

Wandered away another time and ended up in the high school library.

At Christmas time, he ran in my in-laws' house and promptly marked their Christmas tree. In case you don't know what that means, it means he lifted his leg and peed on it.

The first time I took him home to visit my parents, he immediately caught my younger brother's prized fair-winning chicken and brought it to me. DOA, naturally.

We cleverly trained him to bring us the Sunday paper to save us from going outside. He took this task to heart so well, he soon was collecting all the neighbors' papers too. And bringing them home. So much for that job.

"Is that your dog?" she asked. Her husband stood quietly beside her.

We had just walked out of the family cabin at Dillon Beach in Northern California and were on our way for a beach walk. Jeff was walking with us.

"Yes, his name is Jeff," I answered.

In the meantime, Jeff ran up to her like they were the best of friends. She got down on her knees and started petting him with enthusiastic affection.

Her husband seemed less moved.

"I didn't know his name was Jeff, but let me tell you how we know him."

Several months prior, Molly and I had left Jeff with her parents. We were on a trip that did not allow dogs.

During the time Jeff was under their care, Reese and Marge, my wife's parents, decided to spend some time at the beach cabin. Naturally, they took Jeff with them. He was pretty likable, so that wasn't a problem.

They soon developed a routine together. Morning walks to the beach, lay around the house, chew on a bone, and then a last-minute outside pee before bed.

Except one night, Jeff didn't return. Reese waited, called many times, and finally gave up and he and Marge went to bed. I suspect he was not very happy but wasn't going to go hunting for somebody else's dog at night.

In the morning, Jeff was waiting at the door for them just like nothing had happened. Things got back to normal, and in a couple days, they went home. With Jeff.

Well, it turns out this is what Jeff did that night.

After he peed and most likely did a bit of territorial marking around the neighborhood, he returned to find himself locked out. Like most dogs, he had learned to push on a door to see if it would open. I'm sure he did that before moving on.

He started checking doors of nearby cabins. In those days, doors were rarely locked and not all even closed well. Safe and simple times.

Finally, he found one that opened for him. In he went and searched the cabin. He soon found the bedroom. It had a big bed too, probably king-sized. The couple was asleep. Apparently, to Jeff, it looked comfortable and enticing.

Jeff jumped onto the bed and made himself comfortable for the night. The couple was terrified at first. They didn't know what to do. Who would?

Finally, the husband suggested they do nothing. Jeff was already asleep, and it was a big bed anyway.

They didn't sleep well that night, but I suspect the dog did, knowing him. In the morning, he left through the door that he had opened the night before.

Until they saw him with us, the couple had no idea where he came from or who he belonged to.

Of course, we had no idea that had happened and Jeff wasn't telling.

FOR MANY YEARS, Jeff had a companion. Another dog, Missy.

She was a Golden Retriever, and just the sweetest dog ever. Missy didn't have Jeff's personality. She didn't do contrary things.

At some point somebody convinced us to breed Missy and let her have a litter of puppies. We found another Golden to be the father. Of course, much to Jeff's disappointment.

Golden Retriever puppies, by the way, are among the cutest things on earth. It's pretty easy to find takers when they are old enough. Usually around eight weeks old. Anyway, by then even a good mother, like Missy, gets tired of the whole thing.

We kept them in a big box in the kitchen. Missy could come and go as she wanted. Jeff would come to look in, usually in disbelief.

When the puppies got a bit older, sometime after their eyes opened, they got adventurous. Finally, their adventures took them over the side of the box.

This, of course, was going to lead to all sorts of problems. Including some unspeakable messes.

One day, they found Jeff asleep on the floor. He was stretched out on his side. To the puppies, it looked like a new mother to suckle. They went at it. Instantly, Jeff was up and gone. Highly insulted. Out the dog door. They had no idea where he went, well, because dog doors were something they didn't know about or understand.

I didn't think much about it. We had a fenced yard. He would return eventually.

Except he didn't.

After a day, I started searching the neighborhood, walking, driving, talking. No luck. After several days, I called the local radio station. In those days, they would put a notice on the air. I told them about a missing Lab, and left my name and number.

It wasn't long before somebody called.

Yes, a male black Lab had shown up a couple days prior. He just moved right in. Jeff seemed to have that ability, I was beginning to notice.

I picked him up. It was about a quarter mile from our house. Jeff acted as though nothing had happened. He was pretty good at acting that way.

I added higher sides to the puppy box. They stayed in. Jeff stayed home.

21

My New Ice Axe

NOTE: This story includes Jeff, the dog. Eventually. And my new ice axe. Eventually.

As soon as I got a driver's license, a car and, of course, enough money, I started driving to the trailheads of all the big mountains in Southern California.

I wanted to climb all of them. Although they are not climbs technically. They are long, hard hikes. On a trail.

Mt. San Antonio, Mt. San Jacinto, Cucamonga Peak, Telegraph Peak. Others too.

I only missed one peak, Mt. Gregorio.

Rob and I camped part way up the night before our intended climb of Mt. Gregorio. A huge thunder and lightning storm came through that night. Very little rain, but an enormous amount of thunder and lightning.

As we got high on the mountain the next morning, near timberline, we saw smoke. Soon we saw the fire, about 5 acres smoldering through some high-elevation brush.

We didn't have tools for firefighting, but the fire was moving slowly, so we thought we could do something. And should do something. It didn't look that dangerous.

Using our boots to stomp out some flames and kick dirt over other flames, we started working our way around the fire. Soon a small plane flew overhead and then disappeared. Before long, it was back and flew near us. Somebody reached out the window and dropped a shovel and a Pulaski. That's a firefighting tool that's half axe and half hoe. Pretty standard for firefighters. Now we could really help control the fire.

After a couple hours a backcountry firefighting crew appeared and took over our task. By then the day was gone, and so were our hopes of climbing the mountain.

The crew leader told us to check in at the temporary fire office on our way home. They would take our information, names, addresses, etc. and send us each a check for our time fighting the fire.

Some time later I got a check from the US Forest Service for $13.44. I wasn't going to do that for a living.

I MOVED to Northern California to attend the university. University of California at Davis. To the east were the magnificent mountains of the Sierra Nevada. It was full of high peaks. Some had trails to their summits like I was used to, but most didn't. The trails would only go so far. Then it was a scramble through the rocks, boulders, snow and maybe ice.

There were glaciers, small and large. Glaciers represented a danger I didn't feel comfortable about, especially alone. They were slick and had crevasses to fall into and maybe never get out of again. Maybe fatal.

The snow fields seemed different. Almost friendly. Anyway, snow is soft.

So, I convinced myself I needed an ice axe and bought one.

In those days they had wooden handles. Maybe ash, like a baseball bat. The handle could be 2 to 3 feet long. At the end was a miniature shovel for digging in the snow. Opposite it, on the same end, was a long curved spike thing for grabbing snow or ice.

I self-educated myself on its use. Probably not entirely wise. And then I was ready to try it on a snow field. Or so I thought.

My wife and I had a favorite camping spot high on one of the passes. It was pretty remote and rarely had much use. Above it, about a 45-minute hike away, was a peak with a large north-facing snowfield. The snow was usually there year around.

Below the snow field were rocks. Smaller ones lower down and bigger ones above where the snow started. It was easy walking over the jagged rocks and the lower part of the snow field. Soon it got steeper and steeper. Finally, I had to start cutting out steps for traction.

All this time, Jeff was walking with me. That's my dog. He went most places with me. Usually alongside. He could grip the steep snow with his paws and nails.

I kept climbing and cutting steps. It kept getting steeper and steeper. I began to be a little more careful. I didn't want to slip and fall down the steep slope. Jeff was content to walk beside me. He didn't need help. Finally, we reached the top of the snow field. It was really steep at that point. There was a small melted-out ledge at the top. We stayed there for a while. The view was great, but soon it was time to go back.

I started down and called my dog. He was not about to leave the ledge. I asked, I pleaded, I yelled. He wouldn't budge. Finally, I climbed back up to him with the intention of giving him a push off. I knew he could negotiate the steep snow once he got started.

In the process of pulling him off the ledge, I lost my footing and was suddenly falling down the snow. And fast, very fast.

During my self-education of the use of my ice axe, I learned about stopping yourself during a fall using your axe. Basically, you dig the long pointed curved part into the snow. It's called self-arrest.

Naturally, reading about it and doing it are two different things. Learning in the middle of a fall is just a ridiculous idea, as I learned.

By the time I got control of the ice axe, I was moving really fast and a bit spooked. I dug the pointed end in the snow hard. Too hard as it turned out. The almost sudden stop flipped me over. A couple of times and sent my ice axe flying.

Suddenly, I was flying down the snowfield with the jagged rocks appearing quickly. Really quickly. "I'm going to tear my belly open on the rocks. Eviscerate myself", I thought. The snow had numbed my belly. I couldn't feel it anymore.

Finally, I was on the rocks and stopped. They were too rough and jagged to slide over. I couldn't feel my belly. Slowly and definitely with hesitation, I ran my hand over my belly, wondering just how mangled I was. My stomach area was completely numb. I ran my hand over it. I discovered my clothes were intact. There wasn't any damage to my clothes, or more importantly, to me.

I looked around for my dog but couldn't find him. Finally, I looked up. He was still on the ledge at the top of the snow field. Now he was probably convinced to never come down.

I definitely needed a new approach. A new strategy.

I had to go back to our camp. And I had to leave my dog on the ledge at the top of the snow. Not forever, but for now. And I had to think about a rescue plan.

While walking back, I decided to collect all the rope I could find and tie the pieces together. I would then reclimb the snowfield, tie the rope to my dog, hike down to a safe spot lower down, and pull him over the ledge.

Back at camp I found enough rope pieces and then tied them together. It seemed like enough length.

Soon, it quickly clouded up and a thunder and lightning storm started. I couldn't go anywhere. It was too dangerous. I was really worried Jeff might be killed by lightning. I stayed in the tent with my wife.

The storm disappeared as quickly as it started. I began hiking back to the snowfield and Jeff. I was really worried about what awaited me. I hiked as fast as I could. Soon, I was there. And there was Jeff, waiting at the top of the snowfield alive. On his ledge.

I had to recut all my steps. The rain had washed them out. At this point, I was getting pretty good with my axe. Maybe not with the self-arrest part, yet.

It didn't take long to get to the ledge and my dog. Of course, he was delighted I had returned. I was happy too, but the feeling was mixed with some anger about his stubbornness. I put the rope around his neck and started down. Carefully. I let the rope fall away ahead of me. It ended where the snow was not so steep and I wouldn't fall. When I got there, I set myself up and pulled Jeff over the ledge.

Once he was over the ledge, he realized it was safe, and he came the rest of the way on his own. I gathered the rope and we walked home just as the sun was setting.

22

———

Nameless In Hamlet

IT WAS first thing in the morning, but I was already in my small office. The phone rang. I answered it.

"I would like you to look at my animals. I just moved to this area and just bought them and would like health checks or whatever you do," she stated.

"Are they farm animals?" I asked.

"Yes, two donkeys and three goats," she answered.

"Great. Where do you live?" I asked. I have always had a soft spot for donkeys.

She named a very remote road in a distant part of our county.

"I don't get down that way very often. Maybe once or twice a year. Could you bring them to me?" I asked.

"Yes. I have a pickup with sides. They'll all load for me."

We agreed on a date and time about a week or so away. I

didn't bother to get any more information from her. She didn't seem like the kind that would back out.

She arrived on time with her animals in an older pickup with wooden sides that she had hammered together. It was one of those older trucks that might not run well, but would probably run forever. Anyway, that's all she needed.

My assistant, Dan, helped her unload her animals.

I examined each donkey and goat. Dan did some lab work on each. I ended up giving them basic vaccinations, deworming medication and addressed a few minor medical problems. She was curious about their nutritional needs in our area, especially since she came from a different climate. We also discussed that.

After we were finished, I returned to my office to make notes and prepare a bill. After they loaded the animals, she met me in my office.

"I need some information for my records," I said.

"Like what?" she asked.

"Your name and address," I answered. I already had her telephone number from the original phone call.

"I can't give you my name."

I was perplexed. "Why, what's going on?" I asked.

Then she explained. She was hiding, hiding from a very dangerous ex-husband. She was scared. She was really scared. I didn't ask what had happened, but I respected her fear and the situation. Fortunately, she picked a great place to hide out. And she wasn't a criminal either.

For the next several years she called a couple times a year for an appointment. Maybe more often, as she kept getting more animals, goats especially. The pickup kept getting more crowded. But it kept running.

She always seemed happy, and I assumed things were going well.

One day about five years later, she called and told me she was moving. I was surprised. It turned out she had a new boyfriend. She felt safe with him, and they were moving to another state. She had called because she needed health certificates on her animals to move to that state.

This time I agreed to come to her place as she had too many animals to bring to me.

"However," I said, "I need a name and address for the health certificate. This is official state business."

This time she agreed. Her life had completely changed and she felt comfortable and safe.

23

———————

My Very Brief And Unremarkable Intercollegiate Sports Career

As a kid I loved sports. As a kid I was a pretty terrible athlete. I wasn't strong or fast. And I was pretty much without coordination.

I was the kind of kid who would sit at the far end of the bench on my Little League team. If we were so far behind or winning overwhelmingly, I was allowed to play. Only right field, where fewer balls are hit.

If I was in and playing right field, I really didn't want a ball hit to me. First, I would have to figure out where it was headed. Then decide if I could catch it. Catching it was a big problem. I might miss it. It might hit me in the face. And there would go my two new teeth.

In local neighborhood pickup basketball games, I always got picked near the end. Usually only me, and some seriously uncoordinated kids would be left. At least I got picked before them.

As I grew older, I did get bigger and a lot stronger. The coordination part never got that good. It lagged noticeably.

Later I went off to college. It was a small school at that time, and some sports were pretty much open to everyone. Track and Field was one of those. It was pretty basic then. Just a running area. No special shoes in those days. Places to jump or vault. Things to throw, like the shot put, discus, javelin.

I decided to try the shot put. I was getting a lot stronger. Also, erroneously, I thought it didn't require exceptional athletic skills. The coaches didn't care; I wasn't taking up space. There weren't any scholarships to abscond with in those days, so that wasn't a concern either.

I practiced every day with a couple of friends with similar delusions. Truthfully, I didn't learn much or improve much. The coaches were too busy with those who looked at least a little promising.

Finally, after a couple months of training, the first track meet was nearing.

It was at a nearby college and was to include athletes from other colleges as well. Pretty exciting for me.

We met on campus first thing Saturday morning for breakfast. A special breakfast to power us through our events later. For me it was really special. We had steak. Steak for breakfast! That was unheard of in my life, where wieners were considered an important meat. Regardless, I loved it.

I didn't have an official uniform either. I had to wear my practice shorts and top. Pretty dingy.

The shot put area had the circle you threw from and then a series of concentric rings in the dirt beyond. They started at 30 feet and went up to about 60 feet in 10-foot intervals. Nobody threw to 60, or even 50 feet. You could win with a throw in the mid-forties in those days and at those track meets.

I wasn't good at all, as it turns out. I couldn't throw to the first ring. The 30-foot ring. Everybody was better than me. I also noticed they were a lot bigger and stronger. I also figured out this was not going to be my thing.

On Monday, I told the coach I was quitting because my grades were suffering and I hoped to become a veterinarian. And that I needed to concentrate on that.

He thanked me for telling him, but was too nice to let on he really knew the reason.

24

———

S leep Next To The Wall

WE WERE WORKING at a ranch that brought kids in from all over the country. Mainly from the East Coast. It was at the edge of the mountains just east of Lake Tahoe. I was in charge of the horses. My wife helped cook.

The horses were for the kids, teenagers, to ride around the mountains above us. I would lead them, of course.

It didn't require high-level skills to do this. And at that moment that was something I did not have. I did know how to care for the horses, however.

The horses came in all sizes, shapes and colors. Some in mixed colors. But they had one thing in common, they were what we called "bombproof." That meant they were quiet, gentle, and unexcitable. I'm tempted to throw in thoughtful too, but maybe that doesn't apply to horses. These characteristics were essential since these kids didn't know anything about horses. They probably couldn't distinguish between a horse and a rocket ship.

Something else, or I should say someone else, who really helped was John. John Nevers. He was an elder member of the Washoe Tribe. Actually, he was more than that, he was one of their chiefs.

John was born around 1900, in the area of Washoe Valley. At that time and during his younger years, his family migrated with the seasons. Moving from low areas in the winter to high mountains in the summer. Always looking for food. He had grown up with horses and would help me with the ranch horses.

John was short, very stocky, and walked with a rocking motion, which I realize can happen as you age. He seemed to have a very large head, which could manufacture an enormous smile. One of his prevailing traits, smiling.

One day while I was grooming or saddling, or something, John came to me with a plan. He had found a neighbor with a young, and as it were, untested stallion. Actually, untested meant he had never bred a mare before. John wanted to take a mare we had to him for breeding and maybe have a foal eleven months later. He was really interested because our mare was an Appaloosa, a Native American breed, and so was the young stallion.

That evening, after dinner when we had free time, we loaded the mare in our horse trailer and drove to the stallion.

We arrived, unloaded the mare, and walked her to the young stallion. She was obviously receptive. That was essential, and probably fortunate. Naturally, he was. But then stallions always are.

We put them together. The mare stood quietly and agreeably. He approached, somewhat cluelessly, tried to mount, and fell off. Obviously, this was going to take some learning. He tried it two more times, falling off each time.

At that point John had decided that was the funniest thing he had ever seen. I have since learned that most Native Americans have a wonderful, if not weird, sense of humor. And I'm sure it has led them through unimaginable travail over the years.

Finally, John came up to me, put his hand on my shoulder and said, "Be sure you sleep by the wall tonight. I don't want you falling off".

Then he just kept laughing.

I'm not sure we ever got the mare bred that evening.

25

———

S ignatures

IT'S BEEN a long time now, over 50 years. Most people can't remember, or don't want to, or were not alive. But fuel was really cheap. Gasoline, diesel, airplane fuel. Gas for your car was 20 cents a gallon. Maybe less. Flying was cheap, too.

Of course, it all changed after the Middle East countries realized they could boot us out and take over our oil companies. But that's another subject.

During that time, several more advanced Asian countries decided to upgrade their dairy industries. They decided they would do that by importing good dairy cattle from the US. Plus, it was cheap to fly in those days.

South Korea was especially interested. And especially since a dairy cow or heifer could be flown to them for about $200 each. That's about what they were worth, too. So, if they picked out good cows and imported them, after several years they would really improve their dairy herds. And that's what they did, until the Middle Eastern countries dramatically changed things. First, by withholding oil and then selling it for a much higher price. Much, much higher.

Monte had an excellent herd of Jersey cows. Small herd, but with the best of genetics. He had worked at it for years and had extra animals to sell. So, he and the South Koreans made a deal.

It turns out, South Korea had very difficult health requirements. They wanted healthy cattle, of course. That was natural. But they also wanted lots of testing. And that's where I came in. They wanted blood tests for things we didn't have, for things that didn't mean anything, and some where no tests existed.

I worked through it. Slowly, though. It took months. It took lots of phone calls with federal vets and assorted vet labs all over the country. The paper pile kept getting thicker and thicker. But eventually, we worked it all out.

Finally, I had all the results from all the blood tests and my examinations for each cow. I carefully wrote everything up. It was an enormous effort, and I tried to be precise. This was a big deal for Monte, South Korea, and me. For me, because I wanted to get paid for this effort and failure wasn't going to reward me.

Once completed, I had to send the paperwork to the federal vets for final approval. I was ready. When I finally signed the papers, I was very careful so the signature could be read by anybody. It was sort of like my seventh-grade signature, very precise. Not like the one that had evolved over the recent years that nobody could read but did understand it was mine.

I sent it off, certified mail. I didn't want to take any chances of losing the paperwork, and I wanted it delivered quickly so the cattle could get on an airplane and fly to South Korea.

It happened very quickly, maybe after a day or day and a half. I got a call from the federal vet's office. They were upset about my signature. They were sure someone had forged my signature. I guess because they could read it.

It took me awhile, but I finally convinced them it was indeed me. Me being very careful and precise.

After that, I went back to my illegible signature for everything and forever since.

26

———

D enny And The Neighborhood

I'M sure you've never heard of Denny. That's not surprising either. It's a hamlet located along the New River on the southern edge of the Trinity Alps. It's in northwestern California. A very remote spot.

Originally, it got started as a gold rush town. Now it's for recluses and a major Forest Service station.

About the neighborhood.

It includes Willow Creek, which is a real town. And, the Hoopa Valley Reservation, which, of course, is Native American. And a lot of people scattered about the hills and little valleys. That's largely where the recluses come in.

Many years ago, when I was not only younger, but really younger, I decided all the folks up there and their animals needed a vet every once in a while. I was living and working about one or two hours away. Probably closer to two hours of hard driving. Not very convenient for regular visits.

So, I arranged, through word of mouth, to travel through those areas every two weeks. I would care for whatever farm animals needed me. Mainly cattle and horses. Also mules, as it turned out. There was only one requirement: the animals had to be caught up in some manner. That was not usually a problem except on the reservation which seemed to have a different definition of animal containment. But we did the best we could. Some were really good with a lariat. So, in the end, an animal at the end of a rope and tied to a tree was the containment.

I was making a list of appointments for an upcoming round of visits when I got a call from a reservation member with a few cows. It turned out one of his cows had calved a couple, maybe three days before. She had not passed her placenta, or what we commonly called an afterbirth. That could be a problem after several days, as infection could set in and be dangerous or even fatal. I put him and his cow on my list.

When I arrived, I found the cow and her calf in a small, fragile wood corral. The cow wasn't doing well, but didn't seem near death either. I roped her and tied her up. She was pretty weak, so that wasn't too difficult. As I examined her and the mass hanging out of her, I realized it was not a placenta. It was her uterus now shriveling and turning brown. She should have been dead by now. That's just too big of an organ to be hanging outside a cow. Or any other animal, for that matter.

There was no way to clean the uterus up and put it back in the cow. It appeared to be dying anyway. The only option was to amputate it. Actually, that's not that hard to do. He found me an old inner tube and I cut a long, narrow strip from it. I soaked it in a disinfectant. Then I wrapped it around the uterus as tight as I could and as near the cow's body as I could. That would strangle off the remaining blood supply. Then I tied the ends. I cut off the uterus and stuffed the stump back in the cow.

Surprisingly, she survived although that ended her reproductive career.

The Forest Service station in Denny is a jump-off point into the Trinity Alps. The Alps is an enormous wilderness area, full of wildlife. It's very popular with hunters, fishermen and backpackers.

The Forest Service kept a team of about 20 mules and used them to service the backcountry. Trail maintenance and repair, firefighting and whatever else was needed. The mules were of excellent stock and purchased from a farmer in Missouri. The mules were used for 20 years or so and then sold to the locals. A lot of people used old Forest Service mules for backcountry purposes. Like hunting and fishing. Some could be ridden, most were used for packing supplies.

I would occasionally care for these older mules on my trips to the area.

The one I looked at seemed very, very old. His owner had some concerns, and rightfully so, considering his age, which I could not even determine. I took care of him the best I could, but really wondered how old this mule was. I did remember the mule's name, Timothy, maybe somebody in Denny would know.

Denny was the end of the line every time. I never arrived until late and was always hungry.

Bill Jackson, the local sheriff deputy, and his wife, Melba, were always waiting for me.

Bill had the lonesome job of covering all those rural

roads and country to maintain the peace. Their meals were a backcountry delight.

One time it was mountain lion stew. A little bit like chewy chicken, but quite good.

Another time, a bear roast. This bear had fattened on fall acorns. It was as delicious as an excellent pork roast with a little mountain thrown in.

I figured Bill would know about the mule Timothy. Years before, he had worked for the Forest Service and was in charge of the mules.

"I remember that mule," he replied when I asked. "We picked him up just as the war started."

That was WWII, as it turns out, and that meant Timothy was 45 years old. Maybe more.

I made these visits for several more years and then moved on to Oregon.

Not long after I moved, Bill was shot and killed by a criminal he was attempting to capture. It was a tragic end to a wonderful man and a wonderful host.

27

———

Looking Around

Wilson was older, much older. More like where I am now, I think. But then I was much younger. Getting up and down. Never gave it any thought. Bending over for work, the same.

He had an old horse, probably 30, maybe more. He had had the horse since they were both much younger. This was his, not belonging to a previous wife, girlfriend or daughter. I was never sure what he did with the horse. Maybe hunting. That wasn't uncommon in our country. But I wasn't sure. It didn't matter anyway.

The old horse had recurrent feet problems, usually abscesses. That's what I thought was going on at the time. It was very lame, almost unable to stand on its front right foot.

It was going to take a bunch of special tools and some bandage material to get the job done.

I laid all the tools on the ground before me and beside

the hoof. Hoof trimmers, hoof testers, a couple of different hoof knives and a bandage. The ground was my work bench.

I picked up the hoof, carefully cleaned it out, and applied the pressure with the testers until I found the tender area. I then used the knives and trimmers to expose and drain the abscess. It drained easily because it was under pressure inside the hoof. That's why it hurt so much.

As I was finishing the bandaging, I noticed Wilson looking over my arrangement of tools. The tools I had placed for convenience and easy access.

Finally, he said, "I know exactly what you are doing."

"What do you mean?" I asked, unsure of where this was going.

"When I get down on the ground to do something, I don't get up until I look around to see what else there is to do down there. That's why you scattered all your tools at your feet."

Actually, at the time, I didn't know what he was talking about. Now, many years later, I do. Getting down and back up gets harder and finally is really hard as you get older.

28

S leeping With Bears

"MOLLY, MOLLY, MOLLY," he screamed. It was our young son, Matt.

The dog was barking too. Bingo, our sweet black Lab.

Mothers, fathers too, are really fast when their kids start screaming for help.

But probably mothers are faster.

We were enjoying the last light on a warm summer evening. Sitting outside the house, having a drink of celebration.

Our son wanted to spend the night in a tent camped along the small river about 150 feet from our house. I sent up the tent. He had a sleeping bag, pad and who knows what else. Most importantly, he had his dog, Bingo. He was set, we all thought.

Looking back, that was pretty brave for a kid not even ten.

Matt and Bingo had settled in the tent. He had a couple of books to read or look through. The dog snuggled up close. They were comfortable. They were set for the night.

The tent was set up under a large spruce tree, maybe 6 feet in diameter. Big by any standards. The river was several feet away on the other side. A steep bank led down to the water, maybe 5 or 6 feet down.

The quiet water sounds should make for nice sleeping, we all thought.

Unknown to all of us, a bear was wandering along the river going in and out of the water looking for food. Apparently, the bear decided its search wasn't working. It decided to climb up the steep bank and ended up beside the tent and shook off. It was pretty wet from wading in the river. It shook off right beside the tent and showered it with water.

Matt and Bingo had no idea what was going on. Of course, the bear had no idea what the tent was and what it might contain. It walked slowly to the front of the tent. Matt was confused as the bear walked to the front of the tent. Bingo was on high alert. Then the bear looked in the tent as Matt sat up and Bingo barked.

I'm not sure who was more startled: the bear, Matt or Bingo. Matt started screaming for his mother; the dog barked a few more times.

The bear?

It ran. I'm not sure where, but it was fast. Startled bears are really fast.

I have often thought if bears tell stories around camp-fires, or wherever they tell stories, that story is probably still being told.

Matt slept in our house that night. So did Bingo. I took the tent down in the morning.

He never asked to sleep along the river again.

29

———

T echnology

SIXTY YEARS of advances in science, or a scientific field like veterinary medicine, end up producing new things that seem more like science fiction than real. Or maybe even more like magic.

An idea appears, then technology is developed. Then the technology is made practical, economical, durable and miniaturized. That means I could put it in my truck and carry it around to farms in all kinds of weather. Even drop it. But probably not run over it with my truck.

My favorite example is the ultrasound. This technology gives rather amazing images inside and outside of animals. We started learning about them maybe in the 1980s. The machines were enormous, the size of Volkswagen bugs. They were only found in big institutions. Portability was not part of the vocabulary.

That was to change, dramatically. But it took some years.

I was always intrigued with new technology, especially if it would give me more information about my patients and help their situation.

Sometime in either the late 80s or early 90s, I bought my first ultrasound. By today's standards it had dubious abilities, but it was still immensely better than no ultrasound. It had a major problem, however. It was enormous. Too big to carry. Bulky and heavy. Maybe it could have been incorporated into an NFL lineman training program.

It had to be wrestled in and out of the back seat of my pickup and loaded into a wheelbarrow. That meant everybody had to have a wheelbarrow. Fortunately, in the country, everybody has wheelbarrows. The real complication was that they were usually used for hauling manure. So, sanitation was a problem. We usually lined the wheelbarrow with feed bags to keep the ultrasound clean.

After a while, I quit using it. It was just too clumsy and hard to move around to be useful.

Soon something amazing happened. Pretty quick, too. But then technology can change like that. The ultrasounds got miniaturized. Suddenly, they became the size of large laptop computers. Durable and weatherproof too.

This process continued. The machines got smaller and better and tougher. Some companies claimed you could now drive over their ultrasound machines. I wasn't about to try that. But I did know vets that drove off with their machine on top of their trucks. They survived the fall.

The protective cases were also amazing. In their protective cases, the ultrasound could survive a drive over. One advertisement showed it surviving a bear attack. I bought that one. You never know. Fortunately, a really good state-of-the-art ultrasound came inside the case.

And then it finally happened. Ultrasounds got really miniaturized. Now they are just a little larger than a cell phone. They can examine all kinds of things and are connected to our cell phone with Bluetooth. You can read the findings on your cell phone screen.

I'm not sure where this goes next. Your watch? Your ring?

30

―――――

The Military Ball

AT THE UNIVERSITY there was an ROTC program. That stands for Reserve Officer Training Corps. It was mandatory for all the guys for two years. After that, if you were interested, you could take an additional two years plus a summer training and be commissioned a second lieutenant when you graduated. Quite a few did the whole program.

I wasn't interested. The military, in any form, did not interest me. I was interested in science and was a reluctant participant in ROTC that included lectures and marching. We were issued uniforms and a rifle. I don't think we ever were allowed to fire the rifle or have ammunition. Looking back at some of the students, I can understand why. It would have been completely dangerous.

Every year ROTC put on a dance called the Military Ball. It was a big deal for the right person. All the guys wore their uniforms. Their dates dressed pretty formally. It wasn't a big deal for me. I had no intention of going.

One year we had a keg of beer the afternoon before the ball. It was at our fraternity house. The guys in ROTC were talking about the dance, their dates, all those types of things. I was just enjoying myself knowing I didn't have to wear my uniform or even get dressed up for the gala.

As the afternoon and drinking wore on, Rich got increasingly drunk. I thought it was pretty funny. He kept getting drunker until he could no longer stand.

It was obvious he couldn't go to the dance. He couldn't even stand, let alone dance.

I was still thinking it was all pretty funny. Then one of the senior members looked at me and said, "Hunter, you're taking his date to the Military Ball."

I was flabbergasted. "Why?" I wanted to know, as this was outrageous.

He looked at me sternly and said in no uncertain terms, "To uphold our reputation. No more questions. Understand?"

I was a new member and really had no standing in the hierarchy, so I agreed.

I did not have my uniform either. It was left in a locker on campus because I wasn't going to the ball. I didn't know who his date was either, or where she lived. The fraternity leaders found out. I put on my shabby suit and went off to her dorm to call for her. With the greatest of trepidation, of course.

I wasn't sure what I was going to say to her. Finally, I decided just to tell the truth. Maybe she would realize I was a victim too. Maybe she would feel sorry for me. As it turns out, that was delusional thinking.

She came down the stairs from her dorm room looking

for Rich. I was the only one in the lobby. In my shabby suit, not in a spiffy military uniform. I reluctantly told her Rich was indisposed and I was now her date.

Outrage would be a great understatement, but I really couldn't blame her.

Finally, she agreed to go with me. I'm not sure why. There wasn't much conversation as I drove us to the ball. In fact, there wasn't any. It got a little better once we were there and we had friends to talk with. We danced a few times, drank some insipid punch, and finally called it quits in less than half an hour.

I took her back to her dorm. There was no good night kiss, or even a thank you. I certainly didn't blame her.

And I didn't even come close to upholding the honor of our fraternity.

31

———

The Proof Is In The Poop

This is about Christmas. Largely.

I've often thought the best thing about Christmas is the kids. Especially the little kids. Especially before they figure out maybe Santa Claus is just a story.

Kids, at maybe two or so, learn what Christmas is about really quickly. I'm not talking about the tree with all its decorations and lights or the wrapped gifts that may or may not accumulate around and beneath it. How that happens depends on each family.

I'm talking about the absolute miracle that this jolly, little, fat guy can travel all over the world on Christmas Eve being hauled in his sleigh by reindeer, of all things.

Also, he has the magic to get down chimneys with a big bag of gifts, usually toys, and get back up the chimney again.

House after house. Even houses without chimneys, but that's another category of magic.

Our very old house had a chimney. It was made of concrete blocks lined with tile. Pretty good for an old house. I learned it wasn't the original chimney, however. I learned that one day insulating the attic to make the house just a tiny bit more winterproof. There were charred timbers from an earlier chimney fire that almost got away and burned everything.

So, we had a chimney, but only big enough for an average-sized possum to get down. Not a fat guy with a big bag of gifts.

Furthermore, it didn't end with a fireplace, but an enclosed metal stove. A lousy place for Santa to end up, especially in the dead of winter when I always had a fire going. But then, he was full of magic.

When the kids are little, they don't question any of this. When the stockings get hung up, they know Santa's coming.

But this doesn't last. The older kids learn first, or at least get suspicious.

Probably it comes from one of their buddies at school with an older sibling who gets it figured out and just has to share it.

I think there is a probationary period in the emergence of this disbelief, however.

It goes from I can't believe it's not true. It's what my parents have always told me, so it must be true.

Then it goes to-well, maybe it isn't true. But I'm not saying a word because I want the presents. This could go on for a while, like several Christmases.

Finally, it just becomes impossible to believe. It's obvious, beyond an average possum, nothing can get down the chimney. It makes sense that the kids figure it out. As they get older, they start to figure out lots of things. But what isn't fair is bragging about their brilliance to a younger sister or brother. I mean, why ruin all that magic before it's necessary?

That's where we were. An older brother starting to figure out the Santa story and not being quiet about it either. I decided I needed to do something decisive.

Two things.

First, it was Christmas Eve day, and the elk were grazing across the road.

Second, that got me thinking about proof. Proof of Santa and what comes not with Santa but with his reindeer.

When nobody was around, I snuck across the road with some plastic bags. I collected the freshest elk poop I could find. Christmas Eve elk poop. I hid it in my vet office refrigerator until later. Not the family one in the kitchen. That would have caused problems on Christmas Eve, heated ones too.

After the kids were in bed, I put up a ladder and spread the elk poop over the roof.

The next morning was pretty much a typical and wonderful Christmas morning. However, there was this emerging attitude from the older sibling. We could sense it.

After all the presents were opened and stockings emptied, I went out to feed. We had horses and sheep. On my way back I looked at the roof covered with elk poop. It was glistening in the frosty morning. Pretty much prefect for Christmas.

I went into the house excited and told the kids they needed to come look at the roof. They followed me out.

"Look what Santa's reindeer left," I exclaimed with as much excitement as I could muster.

They were flabbergasted, excited too. Really excited.

So, in the long run, I bought a few more years of Santa Claus belief with several small bags of elk poop.

32

Trailers, Transmissions And Clandestine Activities

LORRAINE WAS PRETTY LUCKY. She had a husband who made a lot of money. He let her indulge in her hobby. Actually, it was more like a passion. Horses. Lots of horses. More than she really should ever have, or certainly needed.

Mostly the horses were kept away from the house and on the other side of the barn. Actually, several barns. It seemed like there was always a coming and going of horses. Mainly a coming of horses. It's easy to acquire horses. Her husband couldn't really see them unless he made an effort. And he didn't do that.

But he didn't care. He also had his pursuits.

And that's how they carried on. For years.

She didn't bother him for much. Didn't ask for help moving hay, feeding, cleaning stalls, or fixing things. Horses are always breaking things. Or injuring themselves. But she took care of it all, or found someone to do it.

It did take his money, however. Lorraine was pretty careful not to push that area too much. And as long as she was careful, or thrifty enough, he didn't care. She usually was.

One fall, she had procured a new horse and needed to bring it home. This horse was going to push her into the way-too-many horses category. Or maybe well past that limit.

The problem was that it was a Sunday and her husband was home. She would have to sneak the horse home. But since it was Sunday and in the fall, she thought he would be watching the NFL. Lorraine looked it up. Yes, his favorite team was playing. The 49ers. Not only that, they were playing their rivals, the LA Rams. She knew he would be totally distracted. It was perfect.

Once she got the horse home into the back and behind the barns, he would never suspect she had added another horse. Another horse to her far-too-many horse collection.

She brought the horse home and came up the barn side of the circular driveway. She completely avoided driving the pickup and horse trailer by the house. That worked just fine.

Her next plan was to back the trailer up the hill behind one of the barns. Then, she would let the horse out behind the barn. Lorraine knew that if she unloaded the new horse in the driveway, it would start making noise. Whinnying. Calling for other horses. She wanted to avoid that. She wanted everything quiet.

She got the horse trailer positioned to back up the grassy slope to the back of the barn. Like most women horse owners, she was really good at driving a pickup truck and hauling horses in horse trailers. And she was very good at backing the trailers. These seemed to be essential skills, and she had them.

She started backing up the slope. It had been raining. After all, it was now fall, and that's what happens around here. Unfortunately, the rain makes grass very, very slippery. Sometimes you can get stuck on level, wet grass.

Everything started well. Then she got to a steeper part. The wheels started losing traction. Spinning. Her solution was to accelerate more. Not the best idea as it turns out. The wheels just spun more, but stayed in place.

She did more. Same results. Finally, the wheels stopped spinning. The transmission had gone out. Now she was stuck. Furthermore, she had destroyed the transmission in the truck.

Finally, she decided there was only one solution. She had to go to her husband and confess what happened.

It was late in the fourth quarter and the game was tied. This was a very exciting game.

She approached him. Told him the whole story. She confessed everything.

His response was, "I want to watch this game. I don't care what happened. Go get it fixed."

33

Under The Bed

In 1980, sometime in mid-May, Mt. Saint Helens blew up. A massive eruption out the north side. It scattered debris for thousands and thousands of miles to the east.

A day or so later another eruption occurred. The winds were from the east this time, and the ash came our way. To the coast. About a quarter inch of ash covered everything, including the cattle. They looked like they were wearing gray overcoats. After a few days it rained enough to wash the cattle clean and the ash into the soil.

Before all that, maybe for six months, the volcano was acting up. Spewing gases and steam. Finally, it started to bulge out on the north side. That's where it eventually blew out.

There were also earthquakes.

If you're from Southern California, like I am, earthquakes are almost routine. At least as routine as an earthquake can be. If the bedroom, living room or classroom starts shaking, you just shrug and wait it out.

It seemed like rain was a bigger concern in Southern California. And that's hard to believe, I know. When it rained, we were let out of school early. It was called "rainy day session." To us kids, it made rain a scarier thing than earthquakes. I never figured that out, but we did get out of school early. Maybe that was the main thing to all of us.

On the other hand, earthquakes are pretty rare in Oregon and the rest of the Northwest.

Mt Saint Helens, getting ready to erupt, sent out an occasional earthquake that made it all the way to the coast. I'm sure near the mountain a lot more were felt.

We had just gone to bed when an earthquake rolled through our house. It moved from north to south like it was on a journey to somewhere south. The earlier earthquakes just shook the house. They were different.

As quick as possible, our young daughter, Alissa, was in our bedroom. Big eyed and scared. "There's something under my bed," she cried out.

We explained it was an earthquake and there was really nothing under the bed. I even took a flashlight and looked under her bed.

I'm pretty sure she spent the night with us in our bed, which made perfect sense.

34

T he Ritual, Interrupted

IRENE BROUGHT her young male horse from across the river. He was to be neutered. That is, surgically changed from an intact male to what we call a gelding.

That was going to require a short-acting IV anesthesia and surgery. It's all pretty routine for a country vet, or any vet who works with horses.

Neutered male horses generally make much safer and predictable companions. And that's really important for an animal that might eventually weigh a half a ton or more.

Irene had a request, however. She wanted to have the testicles to take home and bury. It seemed to be a ritual that I didn't fully understand. Maybe it had a Native American background. Maybe it was a cult belief. Maybe it had a spiritual basis.

I didn't really care; it was fine with me.

There could be a problem I explained to her. My dog, a Labrador retriever, had long ago decided the leavings from these surgeries were hers. That means there was going to be some competition.

The anesthesia went well and very quickly I had removed the first testicle. It was about grapefruit-sized and, of course, bloody. I told her to get ready to take the testicle.

As I handed it out to her, she hesitated. Maybe it was the size, or the amount of blood. Blood usually doesn't bother most country people. Anyway, she hesitated.

My dog didn't hesitate. Before either of us could react, my dog snatched the testicle out of my hand and started running across the pasture.

Irene was horrified, speechless too. Her hopes for a ritualistic burial were disappearing quickly across the pasture in my dog's mouth and probably soon after in her stomach.

Let's face it, I was stunned too.

But my assistant, Dan, dropped everything he was doing, left the sleeping horse, and sprinted after the dog. He caught her as she stopped and was trying to figure out how to swallow this grapefruit-sized mass.

He retrieved the testicle and brought it back to Irene. Needless to say, she was greatly relieved.

We quickly removed the second testicle and the surgery was over.

This time Irene didn't hesitate too quickly to take the testicle from me. Much to my dog's disappointment, I might add.

35

Adventures With Billy

THIS IS ABOUT MY VERY, very long-time friend, Bill Franklin. And it's about our adventures from the Arctic to the end of South America and central Asia too.

In scientific circles, in biology anyway, he's known as Dr. William Franklin.

Internationally, he's known for his knowledge and wisdom of South American members of the camel family. The wild vicuña and guanaco. These are the progenitors to the domestic alpaca and llama, respectively.

By the way, the other two members of the family are camels. The Bactrian from Asia. The dromedary from Asia and Africa. Both have been domesticated for ages, except for a small wild group of Bactrians that live in the Gobi Desert in Mongolia. More about those later.

. . .

Meeting Up

One of the many jobs I held in college to help finance my education was being a clean up guy at the student cafe. That's where students would get together for snacks, soft drinks, coffee and, of course, visit. I cleaned tables, swept floors, or whatever.

Often working behind the counter and serving was Merry. Attractive and friendly. We talked a lot when times were slow. It turned out her husband was a student and seemed to have a lot of the same interests as I did. Animals, science, outdoors, adventures. We also grew up not far apart in Southern California.

Merry was pretty sure we needed to meet up. She was also pretty sure we would really hit it off.

We didn't meet up for a long time. Life was busy and it seemed our times always conflicted. It just didn't work out.

Until finally it did. One day at the end of both our shifts, Bill appeared. Merry was happy and introduced us. Bill said his hello, as did I, and that was pretty much it.

No magical attraction it seemed. He seemed rather curt, maybe distracted. Maybe arrogant. "So be it," I thought, "such is life."

As it turns out, life had its own plans for us. We accidentally met up the next summer, working at the same ranch for youth. Bill taught science. I ran the horse string and sort of managed the little herd of cows and calves. Our wives cooked.

That same fall, my first job as a country vet was in the same town where Bill was a grad student and doing elk research.

At that point we were well on our way to becoming lifelong friends.

So much for first impressions. Some things just take time.

Elk in the Surf

Bill's master's thesis required darting and sedating elk. The dart, fired from a special rifle, contained a sedation drug. The elk were then measured for certain things. Like weight. Then tagged and released. Releasing just meant staying with them until they were awake and mobile again.

That was a long time ago. About 60 years. The drugs in those days were limited compared to today. In fact, we only had one drug to use. It was a derivative from curare. That's the paralyzing drug from South America. The drug was a bit touchy to use, but Bill had mastered it.

He would stalk the elk, which wasn't too difficult because they were park elk and used to people. He had to estimate the weight for proper dosage. Too much would be fatal. Too little, no effect. And a lost dart and wary elk too.

Once the dart hit the elk and injected the drug, it took 10 minutes or so to take effect. The effect was paralysis of the animal's muscles. Not really sedation as it turns out. In fact, the elk were totally conscious. Totally aware of everything happening. This was always going to be the weak point of using this drug. But at the time there were no other options. And anyway, it was working out.

There was a small herd of elk along the beach in the park, which has now become the Prairie Creek Redwoods State Park. After a short search, we found a bull elk we could sedate. Bill stalked and darted the animal successfully. It went down smoothly. We did the measurements and tagging and then waited for it to recover.

The ocean was nearby, maybe 20-30 yards away. There

were three of us, and we took stations between the elk and the ocean. We did not want it wandering into the ocean.

Of course, that's exactly what happened. The elk awoke and got up completely traumatized from the experience. It headed for the ocean. We tried to stop it by hazing and chasing it back to land. We failed. It was more determined, or frightened, than we had the ability to stop it. Soon it was in the ocean, swimming to who knows where.

Bill was completely beside himself. Losing an elk in the ocean might be the end of his graduate school career. That's what he thought, and naturally so.

He was an excellent swimmer. Had been a lifeguard and played water polo at a high level. Soon, he stripped off most of his clothes and was in the ocean. This is the northern California coast. This ocean is cold. People don't swim in it much. Maybe kids with high metabolisms and can seem immune to cold. For a while anyway.

Bill swam around the elk and chased it back to shore. That was amazing. We were there, ready to chase it further away from the ocean.

It didn't work. The elk ran around us. It skirted us and was back in the ocean. Bill was ready to go back into the ocean. To chase the elk to shore again.

But that was too much and far too dangerous. The two of us held Bill from going back. At that point we were willing, but regretfully, to lose the elk. But absolutely not Bill.

We watched the animal for a long time. Swimming west. Finally, it disappeared from sight.

Later it was found dead along the shore several miles north. A terrible tragedy.

Fortunately, Bill was able to continue with his project but with no more darting of elk along the ocean.

Today's drugs are much safer. They sedate so well, the

animals are not aware of what's happening. That makes a significant difference.

CAR WRECK in the Andes

Actually, we were driving a pickup truck. But we did hit a car. Or it hit us. Or most likely, we hit each other.

Bill was taking us on a long tour through the Altiplano in Peru. Us included Bill, his wife Merry, me and my wife Molly.

Altiplano means "high plain" in Spanish. It's a huge area in the Andes between the very high eastern and western peaks. Don't imagine it as a flat plain, however. It's filled with hills, valleys, lakes, some rivers, and small and not so small towns. And people. That's where the people live. The altitude is 10,000 feet or more, but surprisingly livable. This is where the Inca Empire emerged and lived for centuries. This is where llamas and alpacas came from.

They were domesticated thousands of years ago from wild guanacos and vicuña, respectively. Bill was studying the wild and nearly extinct vicuña further south. He had access to a pickup for his travels and studies. Having access to a pickup was almost unheard of in those days.

There were dirt roads we could follow. They were laced through the Altiplano. Nothing fancy, but kept up well enough for travel. Cars were rare. Pickups were rare. There were freight trucks. Buses too. Always buses. That's how most people got around.

Or they walked. Maybe with their donkeys carrying their goods. A poor man's pickup. Llamas were still used for packing goods, but mainly in the higher areas or where no roads existed. Just trails or routes.

The lack of traffic meant driving was a little casual.

Which means almost recklessly. Easy to do. Nobody out there.

Except this time. The road was narrow. There was a tight curve with little visibility. Suddenly a car was coming, head-on. Bill slammed on the brakes. The other driver did likewise. We hit, but were almost stopped. There was fender damage. Headlight damage. Dented hood. The vehicles were still drivable.

Everybody was fine. The other driver, in a car, was a business man making, well, business calls around the country.

Victor, the driver, seemed intent on establishing the blame. Although it seemed pretty evident we were all at fault.

There were no patrolling police in that part of the world. Even the police in the nearest village had no vehicles.

Victor started by stopping locals who were passing by. Some alone. Some with their families. Some with their donkeys. There weren't many people as it turned out. He gave his version. Which was, of course, that we were reckless and hit his car.

Bill gave his side of the story too. Strong enough that nobody could make a decision.

Finally, another car came by. Victor stopped it, talked to the driver and came back to us. He was going to ride to the next town and bring back the police.

We knew immediately how that was going to work out. For us, anyway. Victor would have the policeman completely convinced of our guilt. And he would have a lot of time to do it.

It was a long time before he returned with a policeman. They had to find another car to ride in. Fortunately, it was a clear, pleasant day. The scenery was spectacular. So, we didn't mind too much.

Finally, they arrived. Naturally, the policeman was strongly convinced of our guilt. After all, he had been listening to Victor for an hour or so. But he didn't know Bill and how Bill could argue a point. It wasn't long before the policeman was confused. He started asking people passing by, just like Victor had done.

At this point, the day was starting to disappear. Finally, Victor took the policeman aside and spoke to him. The policeman then told us we were all going to his office in the next town for paperwork.

Victor took the policeman in his car. We followed in the pickup. It was getting dark as we arrived.

The police office was pretty meager. Bare walls, a single naked light bulb to light it. A small, messy desk. The policeman sat down and started his paperwork. I'm sure that's what he was best at. He asked Victor and Bill lots of questions. He filled out his report and gave us copies. He was fining both parties. We were both guilty, which was probably true. But a businessman and some Americans probably looked like a good way to raise money. We were not to pay him, however. We would make the payment in the next town. It was sort of a county center, I guess.

The fine wasn't that bad but it wasn't in our budget either. We figured we would have to make the best of it.

We left the office and walked outside. Victor was with us. He had spoken only in Spanish before. The whole time. But now in very good English, he said, "I never pay these. Just throw it away."

We didn't throw it away, but we didn't pay it either. And we didn't spend any time in the next town either.

. . .

Big Jim and the Sendero Luminoso

We're still in the Altiplano, but further south. Something's changing, although it's usually subtle. Sometimes not.

Back to that shortly.

It wasn't long before we came across a solo American traveler. That's unusual in the remote world. Jim. Quiet, congenial and tall. Like really tall. Not seven feet, just a bit below, however. We teamed up and let him ride in the back. Just perfect as it worked out.

In those days there was a movement just starting. It was starting in the Altiplano. What we noticed was some hostility, some graffiti. Not much. Not everywhere either. Certain towns only. Those places felt uncomfortable and we moved on quickly.

Later, this movement blossomed into a full-blown brutal and bloody uprising. It went on for years. It was Maoist inspired. A lot of innocent people died. Had we been traveling there ten years later, I'm sure we would have been murdered.

It took the Peruvian government years to capture, convict, or kill the group off. A few hot spots still remain in the backcountry today.

Jim was a perfect find. Of course, he got free transportation and traveling companions. We got a giant of a man. If things seemed a bit hostile, we just had Jim get out and stand by our truck. He was almost twice as tall as most of the tiny Altiplano inhabitants.

Thanks to really tall Jim, we had an uneventful journey.

Ingenuity in the Andes

We were back in Peru. This time Bill was leading an

ecotourist group. I was along to help, maybe help herd people.

The goal was to go to Pampas Galeras in the southern Altiplano of Peru. Bill had done his doctorate research there on the diminutive vicuñas. These are wild members of the camel family. The domestic alpaca is largely derived from the vicuña. This domestication was carried out several thousand years ago.

The animals were disappearing fast due to poor protection and illegal poaching. Bill's earlier research provided essential information for their management and future population growth.

But there was always more to learn. The ecotourists would find new country, adventure and contribute to the research. Plus, they would provide money to fund the whole thing.

We met up in Lima. I was guiding the group around while Bill was meeting with government officials.

He received disturbing news. It was not safe to go to Pampas Galeras. The Maoist uprising, Sendero Luminoso, was active there. Murdering North Americans would just be fun for them.

That, of course, left us with two problems. Where to go? What to do?

And this all had to be figured out quickly.

We ended up going north up the coast, to the city of Trujillo. Then we turned east into the Andes. Rented Volkswagen Bugs were our form of transportation. Surprisingly, they were made in Brazil. These little cars were simple and durable. You could push them by hand if needed.

There was a weak spot. The tires. City tires. Tubeless too.

We finally had a flat. After all, the mountain roads were all gravel. We changed to a spare and traveled to the next town. After asking around, we found a place that fixed flats. They had only hand tools and no compressed air. Just a hand pump. I was pretty sure this wasn't going to work.

They found and fixed the hole in the tire. That was easy. They put it back on the rim. Then they took a small rope and tied a loop in one end. The rope was run around the outside of the tire and through the loop. The rope was tightened until the edges of the tire were compressed against the rim. As much as possible anyway. Then the hand pump was started. That, as fast as possible. One man kept tightening the loop and the other pumping as hard as possible.

I was pretty sure this was ridiculous and done only as a show.

But suddenly, the tire sealed against the rim. The rope was loosened and the tire pumped full. Amazing. Soon we were on the road again.

Finally, the roads ended and we transferred to riding horses and mules. Two days later we were in a remote area. There were mountains, scattered cattle and guanacos. That's what we came for, the guanacos. This was the most northern herd of guanacos in South America. Bill had started studying guanacos as well as vicuñas. So this made sense. And a new population too.

We had a crew of Peruvian cowboys to care for our camp. And to cook. Also along with us was the owner of this enormous ranch, Cesar. And along with Cesar was his very city-born wife. She required special comforts. One mule was assigned to carry the double mattress for Cesar and his wife. It required a very big horse to carry her, well, because the wife was very large herself. I'm not sure anything made this camping trip comfortable for her.

Every day we hiked to the guanaco area and recorded

their activities. We returned later in the afternoon in time for dinner.

One afternoon we returned to find the cowboys roasting a butchered calf over the fire. There was going to be plenty to eat for days. But the question was, where did it come from? There were no fences, no corrals. Nobody had a rifle either. Finally, we asked.

It turned out they chased it off a cliff on horseback. It died from the fall. Then it was butchered and brought back to camp. Definitely a bit brutal, but delicious.

Kids in the Arctic

I've never understood why our wives let us do this. Take three kids to the Arctic. Far into northern Alaska, into the Brooks Range, and then kayak back to Fort Yukon.

Three kids, 8, 9, and 10. Katia, Matt, Shelly. And Katia's arm in a cast. It was broken.

It takes lots of stuff to travel into the Alaskan Arctic and back out. Two folding kayaks, tents, a stove, sleeping bags, clothes for all kinds of weather, and of course, food. A big rifle too, for grizzly bears or whatever.

In those days, air travel was definitely more casual. For instance, just pack your gun and ammo away in your bags. Nobody asked. It was fine. And you could take a lot more stuff. Lots of bags, whatever you needed. Not like today, that's for sure.

We flew from Seattle to Fairbanks and then spent the night in the Fairbanks airport. We usually just piled up all our stuff and slept on it.

The next morning, early, we took a local flight to Fort Yukon. In those days, the plane might be a DC-3 from WWII. Dogs were common. New sled dogs for the family.

Or a refrigerator strapped in a seat next to its new owner. Pretty casual.

The next morning, we were off to somewhere up the Sheenjek River deep in the Brooks Range. We landed on a small lake near the river. It took the float plane two trips to carry all our gear and people.

We spent the rest of the day putting the kayaks together and then carrying everything to the river. The kids weren't big enough to help much.

Spending all day didn't mean very much. There was only daylight that far north in the Arctic. Night wouldn't exist for a couple more months.

We were no sooner afloat when the kids started yelling. There was a moose calf at the water's edge. Cute little thing. But it meant mother was nearby too. And as it turned out, she was on the other side of the river, which was quite narrow at this stage.

Being between a mother moose and her baby is almost as bad as being between a mother grizzly bear and her cubs.

I tried to untie my rifle. We might need it. I couldn't undo my knot. But suddenly, the river quickly moved us past the mother moose and calf. Things were going to be fine.

The trip continued that way too, fortunately. In those days we didn't have cell phones or even satellite phones. Planes rarely flew over that area. If we had an emergency, our only hope in attracting attention would be to start a major forest fire. That also might be difficult. Unlike today, the Arctic was much wetter and didn't burn well.

After about a week or so of long river days, wildlife, mountains, spectacular beauty, and nobody else, we came to the Porcupine River. It's a really big river. Our little kayaks were almost lost in it. We immediately had a problem. Not the size of the river, but extremely hard headwinds blowing against us. We found the winds so strong that we could no longer paddle downriver.

We were stuck and soon running out of food. There were bears, lots of bears. Not the dangerous grizzly bears, but the almost friendly black bears. We could shoot one, butcher it, and feed ourselves for a while. We thought about it. Seriously too. We were down to eating corn meal biscuits for every meal now.

Then, suddenly, a boat appeared traveling up the river. We flagged it down. It was a hunting party, local Natives from Fort Yukon. They were traveling up river to hunt moose. Their boat was full and would be even fuller with a moose when they returned. But they promised us they would come back and get us after taking their yet-to-be killed moose home.

It was an anxious wait. Would they really return? We were pretty sure they would. If for nothing else, for the kids.

He did, the owner of the boat. He brought his wife too. That was probably the best idea ever. The kids, Katia, Matt and Shelly, flocked to her open arms. They needed a mother. And she was willing to be their mother.

Anyway, they had had enough of their fathers at this point.

Tiananmen Square

It was our second time traveling to Mongolia. But for the moment we were stuck in Beijing. Paperwork. In those days China wasn't exactly friendly to tourists. Lots of paperwork and also lots of young army guys with automatic rifles around.

We had traveled to Mongolia before. Bill was interested in studying the wild Bactrian camel. There were few of these left, maybe a thousand or so. All in the Gobi Desert. The Gobi is a vast, high-altitude desert with tall mountains, oases and lots of pretty much nothing. Camel and goat herders lived there. They had been doing it forever.

After spending time there, guided by some helpful Mongolians, we decided it was impossible. Too much country, very few wild camels and fewer roads, if any. Two flat tires would be fatal. Furthermore, the Mongolian biologists did not want to share research possibilities with Westerners. Bill was still intrigued, however.

Bill was a true entrepreneur. He wanted to explore Mongolia more and decided to lead a group of adventurous tourists across Mongolia. Ecotourists. This way he could raise some funds for his research. I was his assistant. This was not an average tourist venture. This was an adventure, maybe even a little dangerous and unpredictable.

While stuck in Beijing, Jim and I decided to visit Tiananmen Square, which, of course, is famous. Famous for a very temporary uprising and then brutal murders. Jim, a friend, had signed up for the trip.

Beijing was not a city of cars in those days. There were cars. But the roads were dominated by bicycles and rickshaws. The rickshaws were converted bicycles. They were the taxis at the time. And there were always a few waiting in front of any hotel. Waiting for customers.

We couldn't speak Chinese. Well, maybe yes, no and thank you. They knew what we meant when we said Tiananmen, however. Soon we were off on the main roads darting between a few cars and lots of bicycles.

It was mid-afternoon when we arrived. Jim and I toured the Square. The rickshaw drivers waited for us. When we finally got back to them, it was almost evening. We were ready to go back to our hotel. It was starting to get dark.

The rickshaws took off. We weren't paying much attention at first. Anyway, it was a confusing city. But soon it seemed we weren't going in the right direction. In fact, we seemed to be going, or being taken, the opposite direction. It was getting darker too.

"What do you think is going on?" I asked Jim. He wasn't sure either.

It kept getting darker. And darker. Soon it was dark, and we were on streets with few lights on their sides. We seemed to be going into total darkness.

At that point we were convinced this was a kidnapping, most likely a robbery and maybe even a murder happening.

Then we noticed lights ahead. It was a small grocery stand. Kind of like a mom-and-pop grocery store. The rickshaws stopped in front of it. They all seemed to know each other. I think now, they were all family. We figured out very quickly what we were supposed to do. Buy things. And we did. Lots of things. Lots of things we certainly didn't need or even know what it was we were buying.

We finally spent all our money and had been happy to do it.

They loaded us back in their rickshaws and took us back to our hotel.

Losing a passport

We were in Mongolia again and leading a group of adventurous tourists. It was time to go home, back to the States. We had been lucky; nobody lost, or hurt, or sick. Or even arrested.

We had one near miss however. One of the older women fell in love with a baby camel. If you haven't ever seen a baby camel, especially a baby Bactrian camel, you wouldn't realize just how cute they are. Like one of those adorable stuffed animals from Germany.

After some negotiations with the Mongolian family that owned the baby camel, she had managed to trade her husband for the camel.

Now this was a bit awkward for many reasons. The main one, of course, was that he wanted no part of it. It took a lot of new negotiations with the woman, her husband, and the Mongolian family to get out of it. We were in the Gobi Desert and more or less at the Mongolian's mercy. Looking back, I suspect things at least that strange had happened before. He was a retired physics professor, so maybe he understood different realities better than the rest of us. Maybe he was married to a different reality.

Bill decided to stay in Mongolia longer. He hadn't given up doing camel research. I was to guide the group home. That seemed pretty simple to me. Most of the people were very experienced travelers. It just shouldn't have been a problem.

Finally back in Ulaanbaatar, we were in the last stages of boarding for our flight to Beijing and then home to the States. We were seated, getting buckled up and the plane was almost ready to move onto the runway for takeoff.

Suddenly, somebody yelled at me. He was almost delirious with concern. "Russ, I've lost my passport."

That didn't make a lot of sense because you couldn't board without it. But, then, strange things can happen. I told him to keep looking. He did. No luck.

I finally got up, found the steward, and explained things. Fortunately, he spoke enough English to understand. He asked me to wait and went to the cabin where the pilot was getting ready to taxi the plane to the runway.

Shortly, the steward returned. He said the captain said we are taking off.

Let Beijing deal with it. That seemed a disaster waiting to happen in my thinking.

What I remembered about the Beijing airport was a lot of young soldiers carrying AK-47 rifles, or something simi-lar. They all looked about 15 years old, way too young to be armed. Hostile if not downright scary. I could see Will locked up. Maybe me too. Or maybe even shot.

This was going to be a long two-hour flight to Beijing.

About a half an hour into the flight, Will yelled out, "I found it."

It had suddenly become a most delightful flight back to China and home.

36

C olor Therapy

YOU NEED to know some things about Stanton.

He was a fabulously creative interior decorator. He did landscaping as well. Decorator or landscaper did not describe his talents. They were superb. But he was really an artist for the indoors and the outdoors.

Stanton had a group of devoted clients. Not customers, it was deeper relationships. Not transactional either. Although money was certainly involved. His clients were all in the ridiculously wealthy category and had multiple homes throughout the country. They flew Stanton in their private planes. Always with his goofy little terrier too. They flew him back again too. With his dog, of course.

Stanton had horses. Arabs. He had had horses forever and gave them the very best of care. What I couldn't provide, they went to the university for more and better care. Always the best, however.

Stanton was also gay. I'm saying without cultural, religious or political judgment. That's just who he was. This makes sense later.

When you have horses forever and, furthermore, take the best of care of them, they live a very long time. Eventually, they begin to develop old horse problems. Regardless of the care.

He now had an older mare, maybe in the 30-year range or even older. She was starting to develop hoof problems. Chronic infections, damage to the hoof and bone in the hoof. I could provide the help, but it became almost ongoing. It would take a long time.

I was there weekly or maybe twice weekly, trimming away infected tissue, cleaning things. Putting on new bandages.

The mare was slowly improving. Things were changing for the better. She was entering a different stage of healing.

We had available then, and still do, an amazing bandage or wrapping material. It was about 3-4 inches wide and several yards long. More importantly, it was flexible and self-adhering. Wonderful stuff and still the standard today.

The bandage wrap could be bought in various colors or designs. Maybe camouflage. Or cute little pictures of horses or some other animal. More too. Sort of bland I thought.

They also came in vivid colors. Those were my favorites. Bright reds, blues, oranges, greens. You name it.

I had taken a couple of bright blue wraps from my truck to use this time. For some reason, I did not mix colors. I have no idea what was behind my thinking about that. After trimming, cleaning and applying a gauze-like pad, I was ready to wrap the hoof with my bright blue bandage.

I stopped and looked at Stanton and said, "I like to use different colored wrap for the different stages of healing."

Of course, I was totally joking.

He looked at me and quickly retorted, "You know, I might be gay, but I'm not stupid."

37

———

C ulting*

*CULTING, the act of creating a cult.

Shelby had a lot of horses. Actually, Shelby had too many horses.

That's not exactly rare, however. More than a few people have found it's easier to collect horses than sell or otherwise get rid of them.

She had a pretty good stallion. I'm not sure where he came from. Maybe she raised him. Sometimes she had more than one stallion. Shelby picked up old, worn-out mares with good bloodlines. Usually, they weren't very expensive. Usually, they were on their way to a horse slaughterhouse in Canada or Mexico. They went for meat prices, as we say. Pretty cheap. She could, if lucky, get a few good foals before the mare got too old and worn out.

Eventually, this resulted in a lot of horses. Stallions, old worn-out mares, and their now countless offspring. Countless offspring waiting for buyers.

The great thing about Shelby was that she always kept her eyes open for possibilities. In this case, it was for making money with horses. Making money with horses is not a business for the weak or those lacking imagination. Most people do not make money with horses.

A tide had turned in the equine business world. Donkeys were gaining favor. The simple, much-maligned donkey was becoming a popular animal. A suburban favorite.

Before that, donkeys were easy to find, and cheap. Like $25 each. Or maybe even free. Like, please take this donkey.

It seemed like overnight it cost $1000, or even more, to buy a donkey.

I should mention donkeys come in three main sizes.

Large, sort of the old standard variety. This size is less common. Except for the wild donkeys that run across several western states. They are subjected to roundups by the BLM and then sold to buyers.

Medium. Sort of a new flavor, or size. Increasingly common.

And finally, the miniatures. These are close to the cutest things on earth. The fuzzy, long-eared babies are indeed the cutest things ever. And that's when you sell them. Cute, fuzzy and irresistible. At one point, their prices got into the thousands.

Eventually, the local donkey supply disappeared. Or largely so. Demand exceeded supply. Basic economics, of course.

Always creative in her business endeavors, Shelby discovered a source of donkeys. The Midwest. There was no donkey fad there. And there were plenty of donkeys running around everywhere. Even better, they were cheap. They may have been soup to nuts in their variety and quality, but that would hardly matter in the new, hot donkey market.

She hired a dealer to buy donkeys for her. The dealer would gather a bunch for her to pick up and bring west.

There was a problem, though. Usually, the donkeys were poorly cared for. Poor feed, not vaccinated, not dewormed. No foot care either. The hoof just wore down or broke off. They were usually considered the extra animal on any property.

That did mean people were anxious to sell them. $100, or $200 for a donkey was a big deal.

At one point, Shelby's dealer had rounded up about 20 or so donkeys for her. They had been brought in from different places. They were being held as a group somewhere. She was ready to get permits or travel papers for them when they started getting sick. It seemed to be a virus turning into pneumonia. Nobody was caring for them, or even cared.

She made a decision. She was going after them, sick or well. Nobody was caring for them and a couple had died already.

Of course, it's not legal to move sick animals across state lines. But she didn't care. Her worry was saving the remaining donkeys.

Quickly, she drove to somewhere in the Midwest where the donkeys were held. She gathered them up. Treated the sick and headed west.

It was not an altogether smooth trip. In spite of treatment, more donkeys got sick. Naturally, too, since they were enclosed together. And the hauling added even more stress.

They started dying. Slowly. Maybe one every 12 hours.

That became a new dilemma. What to do with the dead

donkeys? Her solution was to find an abandoned side road, drive as far down it as possible and leave the dead donkey.

That actually worked pretty well. She deposited about 6 deceased donkeys traveling home.

The rest, the vast majority, made it to her ranch, were isolated, recovered and were fine. In the end, most would have died had they been left.

The string of deceased donkeys left scattered at intervals from the Midwest to the Northwest should have been perfect making for a cult rumor.

We're still waiting.

38

———————

L awrence

I MET Lawrence soon after moving north. We were working in the same country, but doing different things. I was being a vet, driving around the country caring for farm animals. He was trimming and shoeing horses. Soon, we crossed paths and later, often.

It didn't take long before I realized he was immensely talented at trimming, shoeing and handling horses. He also trained horses, and that might have been where his reputation was foremost.

Lawrence wasn't a big man, but tough and wiry. Quiet, too. Didn't talk much, but you would listen when he did. Everything that he said seemed important. Twinkle in his eye, as they say. Quiet, but wonderful humor, which, as a young vet, I could be the butt of. That was ok since I had great respect for him.

In those days, we dewormed horses by passing a tube through their nose to the back of the throat, manipulated them to swallow, and continued the tube to or near the stomach. Then we deposited the medication, which was in a solution. The reason: the medication was too nasty for any horse to consider ingesting by mouth.

This wasn't entirely a safe procedure for humans or horses. Horses were not fond of a tube being passed up their nose. They had, what you might call, formidable weapons to resist the procedure.

First of all, they weigh a thousand pounds, maybe more. They are strong, much stronger than we are. It's not even close.

Second, they can fight back. Rear up, bite, strike forward with their front legs. The latter was the most dangerous. Quick, powerful and very dangerous. Of course, kicking with the back legs is dangerous too, but we were in the front end for this procedure.

We often used a device called a twitch to help control the horses. It was applied to the upper lip just below the nostrils. Surprisingly, it had a subduing effect on the horse. It has been determined that endorphins are produced in the brain by the pressure of the twitch. An acupuncture technique, as it turns out.

It didn't work on all horses, as high levels of fear could override it. But some horses became sedate and wobbly to the point we had to lessen the pressure.

So, using a twitch became pretty standard when deworming with a tube, or tube worming as it was called.

Ed had racehorses, Thoroughbreds. Big horses. He had a stallion and a group of breeding mares, broodmares as they are called. Maybe 10, I don't remember at this point. Assorted young stock growing up and getting ready to race.

He also had a big indoor arena and an outdoor training track. Quite the setup for our area.

In the fall, after the first good frost, Ed contacted me. He wanted the horses tube wormed for winter. That was pretty standard in those days. I would need help, he suggested. That, I realized, was a substantial understatement. The stallion, most likely, would be fine. He was handled a lot. The younger animals could be subdued, they were smaller. But the broodmares were a different situation. They were big and not handled much, if at all.

I didn't have hired help at that point, at least nobody I would put in that situation. Harm's way, as they say. After much thought, I approached Lawrence. I'm not sure what I offered, maybe nothing. Surprisingly, he agreed. Maybe he felt sorry for me, or just loved the challenge of the whole thing.

We met at Ed's ranch one afternoon. Naturally, Ed wasn't there, called away by work needs.

Lawrence and I got to work. Things actually went well. The stallion, young horses. Then came the broodmares. Big, rarely handled, unruly, flighty. Dangerous.

They proved to be just as bad as we had imagined. A battle to catch, a battle to get the twitch on, a battle to pass the tube. It was a long and scary afternoon. But we got it done and nobody got hurt. The horses or us.

As we were cleaning up together, I said to Lawrence, "I can't thank you enough for helping me."

He looked at me and said, "I'm never going to help you do this again."

I was a bit shocked. This was the most talented horseman, handler, trainer I had ever met. If he thought it was that bad, just what was I doing? Why was I taking these chances?

Fortunately, in just a few years a new product came out. It was better, safer and in a paste that could be squirted in the mouth. It changed everything.

Thank goodness.

39

Losing the Evidence

RUDY HAD a nice little herd of Hereford cattle. About 20 cows and a really good bull. He paid a lot of money for his bull and was rewarded with lots of nice calves every spring.

The cattle were a side business. He had a one-man heavy construction company. Usually, he was building or rebuilding logging roads in the mountains.

One morning he found a dead cow. He thought she had been healthy the night before. She was a good one, too. Prime of life. Good condition. Always had raised a big calf.

He called me, concerned about this death. We talked for a while about what all this might mean. A lot of times a single death is just ignored as being part of nature. But Rudy had a small herd, and that meant every loss was important. Plus, he had good cattle and she was an especially good one.

We decided I should come and examine the cow, cut her

open, look for what killed her, and take samples for a lab diagnosis. That process is called a necropsy with animals. With humans it's called an autopsy. Same process, different animals.

By doing all this, we hoped to find out what killed her and then do something to protect the remaining cattle: treat, vaccinate, change the diet, whatever.

Rudy had a very big and deep hole dug next to the cow. After my work he was going to bury her. Deeply.

I proceeded. And I collected samples. Especially from the liver, which seemed diseased. I placed all the samples on a log next to the cow. We rolled the cow into the hole. I threw in some leftover pieces too. Rudy buried the cow.

I turned to gather up the samples, the evidence I was going to send off to the state diagnostic lab. There was Rudy's dog, a Lab, just finishing off the last sample. He ate all of them.

Rudy came over to ask what happened. I told him about the dog eating the samples.

I was perplexed, maybe a bit upset. Rudy just looked at me and said, "If he gets sick, I'll take him to my dog vet to find out what's wrong. Maybe that will tell us what happened to the cow."

I wasn't so sure about that, but the cow was well buried and we weren't going to dig it up.

The other thing I wasn't sure about, and I have had Labs forever, is they don't get sick from eating bad things very often.

But no more cows got sick or died, nor did the dog.

40

———

Barn Parties

IT SEEMS like in those days, which means quite a while ago, everybody had a barn. At least around here. There were dairy barns to house the milking cows, particularly in the winter. Hay barns for all kinds of cattle and other stock. And all sorts of leftover barns, mainly from raising mink and chickens. Those industries had disappeared years ago. Later, special barns were built for horses. They usually had an indoor riding arena to use all winter long.

Most barns can be cleaned up to host a celebration. Host a party. There's always a group to play music. Visiting, dancing and imbibing follows. Late into the night, too.

Rarely were these costume parties, but sometimes, on Halloween.

One year a local horse barn hosted a Halloween costume party. We decided to go in a two-person horse costume. In those days, Helen's in west Portland rented all sorts of costumes. They would even send them to you.

So, we rented one, a two-person horse costume. My wife made me wear the rear end, which she seemed sure was entirely appropriate. We had two youngsters, a boy and a girl. After some thinking, we decided to use a couple of big cardboard boxes that housed some recent vet supplies I had received. We cut out places for their heads and arms to stick out. Then we covered the outside with glue and stuck hay all over the boxes. The kids went as hay bales.

A mom-and-pop horse costume and two bales of hay. Quite the night.

For me, backing up to the barn wall and kicking it, making lots of noise like errant horses will do, was strangely satisfying. This kind of behavior infuriates barn owners. It's disturbing, damaging, and sometimes contagious among horses.

Now this was a straight party. No costumes. But I couldn't resist. Why not add a little flavor to it? I called Helen's again. This time I ordered two gorilla costumes. Also, I borrowed a Lincoln convertible to drive to the party.

We arrived as two gorillas driving a big convertible. Top down. Nobody, even though they were all friends, had a clue who we were.

Quickly, we were dancing with everybody else. Truthfully, most people had no idea what was going on or who we might be.

But it caught up with us. The costumes were heavy and thick. You simply can't dance for long without getting hot. Really hot. I'm thinking gorillas don't dance much.

We finally gave up wearing the costumes. Some of the

other people wore them too, for a while. It turns out there are always folks who want to try out being a gorilla, not many, but a few.

The next day my son was pretty sure he wanted to try one of the costumes. His plan, as it turns out, was to sneak off into the woods and scare people.

We stopped that really quick. Where we live, Bigfoot is a serious legend, plus everybody has a gun. Who wouldn't love to add Bigfoot or a gorilla to their bag list?

41

———————

Early Ventures in Horticulture

Releaf work*

*Not to be confused with relief work.

After a few years working on lots of dairies around here, I realized they couldn't get everything done. The dairymen. Dairies are busy, really busy, places. Something is always going wrong, it seems. Or there's just not enough time to get normal things done.

One area in particular I noticed was the raising of young stock for replacements. Replacements for the cows that got sick, died or just got too old.

Not everybody could raise enough replacement stock.

There was always a market for new cows. It looked like raising young dairy cows could be a good business. I knew a few ranchers that were doing just that and appeared to be successful.

I told my friend Ed about it. He was always looking for another crop to add to his farming enterprise. I had some land on our place he could use. Just a few fencing repairs and it should work.

He did it. Bought about 20 dairy heifers, young dairy cows. He fixed my fences, and brought them over.

They did just fine all spring and summer. Then the grass started to get eaten down. Not growing back either. It was that time of the year, so naturally, the animals started looking around for more to eat. Finally, they found their way through the fence. A weak spot, I guess.

I came home late one day to find them all in our front yard. Eating. Eating everything they could find.

They had stripped all the leaves off the rhododendron bushes around the house. These leaves can be poisonous, but not always. They weren't in this case, as it turned out. Maybe it was the plants. Or maybe there were enough heifers to spread out the possibility of getting sick.

We herded them back and fixed the fence. They were going to need to get extra feed or be moved shortly.

The rhododendrons were naked. No leaves. Not much we could do until new ones grew back. Probably the next spring. Ugly bushes in the meantime.

One day, Alissa, our daughter, asked for some paper and scissors. She got out her crayons and kept herself very busy on some project. She was very intense. For quite a while.

Then she asked for some Scotch tape. It seemed to us she was making quite a mess, but was staying occupied. We left her alone.

Later, she gathered up her project and took it outside.

She was gone for some time. Actually, for a long time. We weren't concerned, we knew she was somewhere in the yard.

Finally, she came back in the house and said, "I fixed everything."

"What are you talking about?" we asked.

"The bushes."

We had no idea what she was talking about. Finally, she led us outside to the rhododendrons.

She had fixed them all right. She had been cutting out leaves from her paper, coloring them green, and taping them to the bushes.

She releafed them.

Sprouting Pencils

In those days everybody used a lot of pencils. Letters, notes, diagrams, scribbling. Who knows what else? Just plain writing was common in years past. People actually wrote letters by hand. The very serious would write a practice letter with a pencil. That's hard to believe today.

We used pencils all the time. They were usually scattered about the house. Though most seemed to be in a coffee cup by the phone in the kitchen. With some paper nearby to write on.

Keeping pencils sharp and useful means you have to sharpen them. Somewhere, maybe on the back porch, was the sharpener. One of those grinder types that had a handle to turn. It also had a container that collected the shavings. It took a little practice. Too much, and it used up too much of the pencil or made the point so sharp it broke when used. Not enough, and there was nothing to write with. A learned skill. Sort of.

Of course, everybody kept using pencils until they were essentially sharpened to nothing. Or were so short nobody could hold them. For some reason there was always a collection of too-short pencils that never got thrown away.

One day, Alissa decided it was time to do something with all our very short, too short, pencils. She was going to solve the problem.

She went through the house and collected the way too-short pencils. Then, she took them outside and cleared an area so it was just dirt. She then planted each pencil. I'm not sure now whether it was point down or eraser down. Regardless, her goal was for the pencils to grow back to their original length.

They would then be harvested and returned to our house.

Well, it didn't really work. Actually, it didn't work at all. As you can imagine.

But in the long run, she has become a very successful gardener of real things like flowers and all kinds of things to eat.

42

———

F reemartins

IF YOU'VE BEEN around cattle, you probably know what this means. If you haven't, you're probably thinking, "What?"

Quickly, this is it.

Cows almost always have a single calf. Twins are rare, but do occur. Rarely more than 2% of the time, however.

Also twinning, as it is called, probably is inherited to some extent.

You would think that twins in cattle would be a good idea. More offspring. More to raise. More to sell. It sure works for other farm animals. Sheep and goats, for sure. Pigs essentially have litters. No problems.

But it is a problem in cattle. If the twins are male and female, the female does not develop properly. Her uterus and ovaries don't develop, or only partially so. Apparently, this occurs because the two placentas fuse, and the tiny amount of male hormone the fetal bull produces finds its way to the other twin. This stunts the female's ovary and uterus development.

The result is the female is sterile. Not good for breeding, only fattening and eating.

It happens about 90% of the time with cattle twins with both sexes. Maybe more. If both twins are of the same sex, they are born normal. So, they are ok. But half of all twin pregnancies will be of mixed sexes.

Hence, they are called freemartins, which is a weird term. But it has been around for centuries. As it turns out, the term comes from the British Isles. Mart is the Gaelic word for cow. So, apparently, freemartin means a cow free of pregnancy. Forever.

Many farmers have learned to identify these animals as they age. They develop uncow-like characteristics. More like a bull/cow mixture.

Vets can diagnose freemartins by rectal palpation or ultrasound. They can do this when they are young. Maybe a year of age or so.

Leo had a big operation. Beef cows and sheep. He was a bit of a wheeler-dealer too, as they say.

Now, this was years ago, before the Middle Eastern countries figured out they could take over oil production for themselves and start charging us a lot of money.

This was also the time when South Korea, maybe Japan too, decided to upgrade their dairy herds. They were going to do this by importing dairy cattle from the US. We had much better cows, so that made sense. And it was cheap to get the cows there. Actually, they weren't adult cows, but young animals, heifers. Fortunately, it only cost $200 to fly a young cow to South Korea.

Furthermore, South Korea was paying good money for the animals.

Leo figured out pretty quickly that he had a way of making lots of money. He hired a couple of cattle dealers to find him dairy heifers to sell to South Korea. Soon, the animals were appearing at his ranch in large numbers.

They required a lot of processing, and that's where I came in. They needed to be examined for health and potential breeding. Then there was disease testing and paperwork. A lot of both, as it turned out.

It seemed as though I was looking at a lot of young cows very quickly.

One day, as I was working through a large number, maybe a hundred or so, I kept finding freemartins. A large number. So many that I questioned myself at first. By the time I finished, I had rejected about 50 heifers as freemartins. Certainly, an unreasonable number.

As I looked back at the group, which we had marked with livestock chalk, I noticed something. They all had green tags in the left ear. This group had previously been looked at and deemed freemartins. Somewhere, some place. Some other vet. These animals should have been culled.

I told Leo my findings and my thoughts. Basically, he rejected what I said.

About six months later, Leo was still doing this, and there was another group of heifers for me to look at. I wasn't paying much attention at first. Soon I was finding a lot of freemartins. Like before. Then I started looking at these heifers. They all had green tags. In their left ears, too. They were the same group from before!

Leo was there. Finally, I stopped and said, "The heifers are still freemartins. They don't grow out of it. But it's your money and I'll keep checking them if you want."

He finally conceded there was a problem.

It wasn't long before the export market fell. He was left with a large herd of young dairy cattle that he couldn't sell. There was no market.

His solution was to start his own dairy. So much for the quick money.

43

———

L eaping Lizards

IN SOUTHERN CALIFORNIA, after the orange trees and before the endless houses, there was lots of open country for kids to roam. It was mainly grassland. From 2-3 feet tall and dry by June. We spent a lot of time in that open country. It was surprisingly full of wildlife, birds in particular. Valley quail too, lots.

There were other animals there. Especially lizards. Lots and lots of what are called fence lizards. Small and almost friendly-looking. Also, a few snakes.

However, there was another lizard. Not as many and much different. They were called alligator lizards. They were big. Up to 2 feet long. For us kids, it seemed more like 4 feet, or maybe a lot more. Scary things. Scary as hell, in fact.

John's parents lived on an old farm that had raised

chickens long ago. Everything was still pretty much in place. Chicken houses, storage sheds and even the egg incubator.

The three of us, John, Jim and I, decided to find quail nests in the grass fields. We could then collect the eggs and incubate them. Of course, we were going to raise the babies, but after that our plans got fuzzy. If they even existed.

We picked a nearby field where we had spotted quail before. Our plan was to come from three directions and flush the mother quail off her nest. It would make finding the nest easier, we thought.

It worked. As we got closer to each other, a quail got spooked and flew between us. It looked like we could figure out where the nest was.

Of course, we were making a lot of noise in the grass and yelling too. That caused another animal to spook. An alligator lizard. And it was a big one! It ran straight towards me in its panic. I saw it coming! It wanted to get away from the commotion and just hide somewhere.

The enormous lizard, that's how I saw it, ran across my shoe and up my pant leg. On the inside. Yes, the inside.

I was terrified, anybody would be. Was it going all the way up? Was it going to end up in my underwear? Just that led to endless and rather horrible possibilities.

Of course, you don't really have time to think about all that. In fact, I didn't think at all. But I did put my hands around the top of my pants before it got as far as my underwear. I held tight, too.

The lizard stopped. Now it had no place to go, at least not up. Although it had stopped, I had no idea what to do next.

Fortunately, the lizard did. It turned around and went back down my pant leg. It quickly disappeared in the grasses.

That was probably just another day for a wild creature like that really big lizard. But it wasn't for me. And still isn't after all these years.

We rethought our quail-raising endeavor. We abandoned the idea.

44

———————

H ome Brewing

LET'S FACE IT, more than a few of us did more drinking before twenty-one than after. At least the rowdy, uninhibited, very stupid type of drinking. The problem was always where to get the alcohol.

It wouldn't take much alcohol either. My buddies, Jim and John, and I could split a 6 pack of very cheap beer and have a real party. But we still had to find it.

For a while, we used a small mom-and-pop store tucked away in the farm country south of town. That required a friend with special connections. In this case, they all went to the same church. The Mormon church, of all things. Finally, the owner got in trouble, and then we were on our own.

That's when we decided to make our own. Brew our own beer, in this case. We needed a place and picked John's house. It was out in the country and had a third story that nobody ever visited. Anyway, John's parents didn't seem to care what we did up there. Maybe they did care, but not after his older brother's mischief and near-criminal activities. I think they considered us refreshingly boring.

Furthermore, we never got caught at anything.

We procured the needed equipment and brewing supplies and started our project. Because it was upstairs, it was plenty warm for brewing. We quickly had a product. Although after tasting it, we weren't sure what we had. It had lots of fizz and alcohol. It only sort of tasted like beer. We were hardly connoisseurs of liquor at that point in life, but we knew alcohol when we tasted it.

Finally, the day came. We declared it ready to bottle up into a bunch of cleaned-up, used quart beer bottles. We didn't have a capping system, so we used corks that we had also collected.

A couple of days later, on the weekend, we set off to the drive-in theater for movies and beer. In the privacy of my car.

After it got dark and the movie started, we pulled out our bottled beer, or whatever it was. By that point, it was warm, but again, at that stage temperature wasn't important to us. Just effect.

I had my bottle squeezed between my legs trying to pry the cork loose. No corkscrew, of course. I finally got it out, but had shaken the beer enough so that it poured out all over my pants and car seat. I knew this would be trouble later. Still, we had a most wonderful, stupid teenage drinking night.

I knew trouble would come later. I started thinking up my story before I ever got home and got in bed.

The trouble would come in the form of my mother. My

father would be off doing things. Anyway, he was a teenage boy once.

My mother was dealing with three teenage sons with no experience raising children let alone teenage boys. This inexperience, well, was greeting her on a daily basis. She was an only child, so she had had no experience with siblings, let alone brothers. This was a real-time learning process for her.

All this meant she was, or tried to be, controlling. It was only partially working, however.

For instance, she would come into my bedroom early, while I was still asleep or feigning so, and pick up my clothes to wash. Supposedly. She was always washing clothes, it seemed to me.

And that's just what she did that morning. My pants were reeking with old beer. I'm more than sure she was outraged. As outraged as a mother could be. Probably rightfully so.

But I had been thinking about this and was pretty much ready. I thought.

I finally got up, dressed and went to the kitchen, where she always was. I tried to be casual. Her stare was beyond hostile. I tried to hold myself together and then told her what happened, or at least my made-up story.

"You must have noticed the beer smell on my pants. The strangest thing happened. We went to the bowling alley. I sat in one of those plastic chairs. Some guy had spilled his beer in it. I sat right in the puddle of beer. I stunk for the rest of the night."

The fire went out in her. She had nothing to say. I'm not sure she really believed me. Maybe she figured that was such a good story she would leave it alone and let things drop.

45

———

Fraternity President

I HAD NEVER THOUGHT about or known much about fraternities. The only thing I thought I knew was they were for the wealthy. That certainly wasn't my background. Mine was working class that had climbed out of the depression.

As it turns out, there were a lot of young guys in my situation. And we were all off to colleges or universities to get educated. That also meant, at the right school, we would be joining fraternities.

That's what happened to me. I found a group of guys that I really liked. We had common backgrounds and interests. Anyway, the dorm food was terrible.

It wasn't an Animal House experience either. Not usually, anyway. Although that might creep in occasionally.

Eventually, I ended up being the president. I'm not sure I

can truly describe that experience, but it was certainly educational.

We were throwing a party one Saturday evening. That meant we could invite girls into the house, which was usually forbidden in those days. Of course, no booze was allowed. At least in the open. But it was there, if you knew who to ask.

We had to be pretty careful, however, because there was always a faculty adviser who was attending as a chaperone. University requirement.

It was going well until one of the assigned secret purveyors of booze, well, imbibed too much booze. And in his generosity, he offered a drink to our faculty chaperone. That definitely crossed the line. That not only ended the party, but we were also put on what they called Social Probation. No more parties or social gatherings for the remainder of the school year.

Not that that really had any impact on our drinking.

ROWLAND WAS A GOOD STUDENT; in fact, he not only graduated with a BS, but also got a PhD in biology, eventually. I am amazed, still today. I was sure his future was prison.

I think all the bad things, or at least questionable things, occurred on the weekend. Like our Social Probation. But there was more.

It was an early Sunday morning. That means 10 or later. Most were in bed, and the place was a Sunday morning mess. There was loud knocking on the door. At that time, the president's room was next to the living room where the front door was located. I got up, put a few clothes on, and went to the door. I was met by a police officer, Rowland, and the Dean of Students.

It turned out Rowland was involved in some mischief in the next town. And I assumed, most certainly, it was exacerbated by alcohol. It was apparently quite the chase, on foot, but the neighboring police finally captured him. Now, here he was before me. All this had happened the night before.

Quite frankly, I didn't have any experience in criminal matters. Actually, what that means is, my friends and I never got caught. But we'll leave that as it is.

I was a bit perplexed. But even more so when the Dean berated me for Rowland's activities. I couldn't, for the life of me, figure out how it was my fault. Being perplexed was probably a good thing because I didn't have anything to say. And I most certainly could have said the wrong thing and made everything worse. So, all I said was, "We'll take care of him."

By then the living room was full of guys waiting to hear Rowland's story. As tired as he was, he obliged. Quite willingly as a matter of fact.

THIS TIME IT WAS SUNDAY, again, but later in the afternoon. A car appeared in front of the fraternity house, not the kind of car students have. It was rather fancy and meant to go fast. There were two guys in the front and Rowland in the back seat. Rowland got out and came in the house to tell us what he had brought back.

It turned out he was hitchhiking back from the Bay Area and got picked up by these two guys. It also turned out they had been robbing liquor stores. The trunk was full of hard liquor. They offered to sell us as much as we wanted. Naturally, we wanted it all but had limited finances. Everybody bought as much as they could and at special prices. Then, the two guys were gone. I think they had to keep moving to keep from being arrested. And maybe to find new markets for their products.

IN THE LONG RUN, everybody finished school and went on to be successful in many, many fields: doctors, lawyers, pharmacists, veterinarians, scientists, business owners, engineers, farmers, ranchers and more.

You would never have known it at the time, however.

46

L

istening to Books, a Dilemma

IT SEEMS like I always read, a lot. All kinds too. Naturally, textbooks which were required. But for pleasure I read history, discovery, adventure, poetry, and endless fiction. Science reading was part of my livelihood, but I really enjoyed science beyond my livelihood.

The problem? Time. At the university, studies usually took a lot of time and energy. Once working, well, working did the same. Actually, more. Mainly because of more responsibilities. Then later, a family and what that entails. Reading became a last thing at night and short too, with the book falling on my face as I just fell asleep.

Surprisingly, one of my big times to read, was during finals at the university. I was fortunate to study quickly for my tests and then have a lot of free time. So, I read. A lot.

During one finals week, I read seven novels. They were a good distraction.

One of my friends, Ed, told me he only listened to books. Audiobooks, they're called. He listened to them all day as he drove around on his job. He ended up, what we would call, well-read, by doing this. He thought I should give it a try. Back then, you needed to put a tape or disc in the slot by your car radio and then broadcast it through your speaker system. Like playing your favorite music, except a book came out. The books were really well done, and still are. Great readers. Now, you can do it with your smartphone and ear buds. Same outcome, however.

So, I gave it a try.

I picked a great book with a great reader, too. Exciting story.

I drove off to make farm calls and turned on the book. It was interesting and I quickly got involved in the story. However, soon I made it to my destination and had important things to do.

This went on for a while, maybe a week. But I found it more and more difficult to stop the book at my destinations. Sometimes I would take a longer way to the farm, or slow down, or just stop along the road. Especially if I got to a really interesting or exciting part.

Then the next step was that I couldn't stop, as it was too interesting or exciting, or both. Then I decided the farm call wasn't that serious and I could do it tomorrow, or whenever. Once I started that approach, I was making a mess out of work and certainly not helping my clients or their animals.

There was only one solution to my addiction. And it was an addiction at this point. Go, as they say, cold turkey.

I did it too, and went back to reading at night and falling asleep with a book on my face.

47

———

M oving On, For Now

THERE WAS A LOT OF TURBULENCE. Turbulence inside her. Turbulence around her. Family, friends, school. Regardless, it was there.

But there was one wonderful thing, and it probably saved her life. Her horse. An older mare. Quiet, gentle, and, maybe most important, non-judgmental. She just did what Shirley asked.

Shirley rode her horse everywhere. She was fearless, Shirley. The mare was trustworthy and willing. It worked just fine for both of them.

There would be an occasion when I had to go find them. A call from her mother. Shirley and the mare would be out riding somewhere. It would be an emergency when the mare had been too willing and ended up cut or injured. Some place they shouldn't have gone or something they shouldn't have tried. Never bad, but needing help. I would find them and care for the mare. And then send them home.

Late one day when I was returning home, I saw her riding to the west. The opposite direction I was coming from. Her saddle bags were full, too. I didn't think much about it. It was summer, and I figured she was out for a longer ride than usual.

As it turned out, that's just what was happening. But even more.

On the east end of town there was a set of old buildings that served as the county fair once a year. Otherwise, it was rarely used. Almost abandoned for most of the year. It had a small showing arena, stands to sit in, and several small barns full of different stalls for all the different fair animals.

Nobody paid much attention to it in the off-season, which was most of the time. There was a caretaker, but he could rarely be found. Maybe he was part-time or had other chores elsewhere.

A few days after I had seen her along the highway, word got out that she was missing. Nobody knew what had happened, except she was gone. So was her mare.

At that point I figured out what I had seen was Shirley running away from home with her horse. What a gutsy thing to do and pretty clever using her horse.

Except that horses are big. They can't be hidden that easily. It was confusing that nobody had spotted them.

Probably nobody was going to kidnap a girl and her horse, so that wasn't a concern. It was more like, where could she have possibly gone?

The mystery finely came to an end when the caretaker showed up and made his rounds at the fairgrounds. Well hidden in a back barn, he discovered Shirley and her trusty mare. She had managed to stay hidden for days.

Of course, at that point her parents were notified. Soon a horse trailer was sent to pick up the mare. They were all home shortly.

I have always hoped her parents were happier to have Shirley and her horse home, than mad at her. Let's face it, it was a pretty courageous feat to pull off.

48

R oping Steers

THESE ARE SPECIAL STEERS, not just any steer. The best are from Mexico, the northern part. The Sonoran Desert. After several centuries of special breeding, which actually means those that survived the vigor of the desert, a special breed evolved. Mexican cattle. Long, lanky. Big-horned, tough, durable.

Every spring Jon would buy a group for practice before heading off to the rodeos. Like any sport, practice is not only important, but essential. This was really true with team roping: two cowboys to rope, two horses to ride, and a steer to capture by the horned head and back legs. Jon always bought a small group to rope so as to not overwork any one of them. Often these cattle were more expensive than our normal breeds that we raise for steaks, roasts, hamburger. A specialty crop, I guess.

There are dealers who specialize in finding these steers, getting them across the border and selling them to countless ropers for practice. To rodeo outfits, too. Probably a semi-big business.

This spring, Jon had bought his normal contingent, maybe 6-8 steers. But things weren't right. These normally thin and lanky steers were even more so. Which is hard to believe.

He called, and we talked about his concerns. I agreed to come look things over. We made an appointment.

The animals were in a small pen next to a runway we could herd them into if needed for handling and treatment. It was needed. They were skinny, had diarrhea, had infected eyes, and some were coughing. They were sick. We herded them into the runway, and I commenced with treating them. Deworming, antibiotics, vitamins, minerals. We also changed their diet. Substantially upgraded it. Better hay and now some grain.

I pretty much covered all the bases.

Some time later, weeks maybe or a month, Jon called back. He had a problem with the steers. I was really concerned. They should be fine at this point.

He got to the point quickly. The steers were fine. The problem was they were too fine. They had all responded to treatment, gotten over their maladies, gained weight and were quite active. Too active, as a matter of fact. They were so fast the horses couldn't catch them. They were outrunning the horses.

We talked about it for a while and then decided to revert back to a less nutritious diet. Starving them into submission, you might say.

It took a couple weeks but it worked.

49

___

T he Land of Fire and Barbecues

THIS TAKES a small geography lesson to get started.

At the end of South America, past the mainland, there are islands. Maybe a thousand or more. I'm not sure. They range from large rocks to the big island, Tierra del Fuego. It's about the size of Switzerland and much bigger than all the others put together.

It's called Tierra del Fuego because early European explorers saw fires and smoke from the island when they first discovered it. Hence, the Island of Fire. As it turns out, it was fires set by the inhabitants to improve grazing land for the wild guanaco. Guanacos are wild llama-like animals that they depended on for food, clothing, shelter and probably anything else they would need. Kind of like a Swiss Army knife of animals.

The terrain varies from almost desert in the north, to

extensive forests, to large mountain-filled glaciers as you go south. There are also countless rivers and lakes.

There is one large river that starts in some small mountains on the westside and flows east to the Atlantic Ocean. It becomes quite large and is appropriately called the Rio Grande.

It's pretty famous among fishermen from North America and Europe. They fish for king or Chinook salmon, brown trout and rainbow trout. It's interesting that none of these are native but introduced from Europe (brown trout) or North America (the salmon and rainbow trout). They have thrived, hence the arrival of world-class fishing.

Argentina owns the east side of the island, and Chile the west side. It's not evenly split. Two-thirds or more belong to Chile. The boundary has been contentious at times and has almost led to war. The last time it was settled by the pope. Things are quiet now.

We were traveling the Chilean side just looking around and fishing the smaller rivers that led to the Rio Grande.

We were me, my son Matt, and his son Mack.

The little rivers had lots of interesting names, such as Rio del Oro, Rio Viento, Rio Miradores, Rio Russfin. They often had different names across the border, so maybe the Pope didn't settle everything.

Not many people live on the Chilean side. There is a small city at the northern end, Porvenir. After that, just a few scattered settlements and fishing lodges. We spent the night in one of the lodges. It was made from metal shipping containers. Not in the deluxe category, for sure.

There are rather poor roads, gravel or worse. That matched unfavorably with the questionable tires on our rented vehicle. Flat tires are inevitable.

It seems there is only one place on the Chilean side to

buy gas or get tires fixed. After Porvenir, anyway. That was at Russfin. It had been an enormous ranch at one time. Sheep ranch. Down there, they are called stations and may have a school, church, store and an airstrip besides the houses, barns and corrals.

Russfin had been sold a number of years before to a large company. They built a sawmill and logged the surrounding forests which were enormous in size. The wood, from a beech-like tree, was sold to Argentina and Chile. It is a beautiful and versatile wood, used for everything from burning for heat to house construction and fine cabinets.

They also had housing, a gas station, and did repair. Including fixing tires. That was a big deal in that country too. There was also a lodge-like facility that could be rented. Food and wine, too. Not advertised, of course.

Naturally, we had a flat. We were a long way from Russfin. We found a rural police station and hoped for guidance. Unfortunately, the officer spoke about as much English as I did Spanish. Then he had a brilliant idea and opened a translation program on his computer. After going back and forth on it, he guided us to Russfin.

Driving carefully, we finally made it there. Our language skills were limited. We weren't part of the lumber business either. So, we weren't sure how to handle everything. But the manager, Roberto, did. He wanted to practice his English. We were set. He gave us lodging, food, wine and had our tire looked at and hopefully fixed.

It turned out it wasn't fixable. It looked like we were stuck. Then Roberto told us he could order a new tire from the mainland and have it sent down. He would have a returning worker bring it.

Nice plan, but it would leave us carless for a day or so.

And we only had so much time. Roberto said he would try to figure something out for us.

Later that evening, after dinner, we were watching TV. The Super Bowl of all things. About 10,000 miles from home and in Spanish.

There was a knock on the door. We opened it and found a man asking if we wanted to rent a car. His friend had one to rent to us. He would take us to his friend. Actually, his friend and his friends were barbecuing in the woods nearby. We needed to follow him.

We needed a car. But this sounded a little suspicious, or even more than a little suspicious. Maybe robbery, or worse. But we followed. As it turns out, his name was Raul, and he was from Easter Island or Rapa Nui as it's called in South America. It's part of Chile, so Raul was Chilean. He spoke good English since having spent time in the US as a student.

It was a winding route and he seemed not sure at times. We weren't sure at all, but for other reasons.

Finally, we got there. There was a 55-gallon metal barrel with a fire in it cooking some meat on a grill on the top. There were several other men. One was cooking, or barbecuing, large chunks of meat. They were all drinking cheap beer.

But mainly they were friendly and wanted to share their food, whatever it was. And their beer. And maybe rent a car to us. We realized this was not a staged robbery at all. And that made the cheap beer quite delicious. They were from different places besides Chile: Argentina and Venezuela. Happy to be there, working and enjoying this meal and friendship.

And the meat. Well, it turns out it was beaver. A big one too. One of them had killed it with a shovel of all things. Butchered it and was now cooking it.

Beavers came to Tierra del Fuego from North America sometime after WWII. They were supposed to be a new crop to harvest. A fur crop. It didn't work out at all. Except for the beavers, who found a new home with no natural predators to worry about. Only men with shovels, I guess.

What does beaver taste like? Not chicken. Not at all. But it wasn't that gamey either. I guess it tasted like beaver. It did go well with the cheap beer, however. So, maybe that's how you should eat it.

For a long time, we visited about all sorts of things. Raul helped us communicate. So did the beer.

We missed the Super Bowl, but as it turns out, a beaver barbecue was much better and more enjoyable. Anyway, there would be another Super Bowl, but probably not a beaver barbecue. For us anyway. And at the end of the earth.

Our new tire came in the next day, unexpectedly, so we did not need to rent the car.

50

L ike Father, Like Son

The phone rang. I answered.

"Hey Doc, I want you to come cut my calves before I take them to the auction."

That means he wanted them neutered into steers.

It was Garret, he had a small herd not far away. Maybe 15-20 cows and a bull. Offspring too.

I rarely went to his place, but that wasn't unusual. Small outfits often had few problems.

I was curious about why he needed me to cut his bull calves. He explained he kept getting poor prices at the auction. For years. Much poorer than other people. Maybe 20 cents a pound less. That was a lot in those days. He sent them as bulls, so the auction had to deal with that problem. Usually, the price difference wasn't that much, however.

I wasn't sure what was going on, but set up an appointment in the near future.

When I arrived, he had all the bull calves caught and ready for me to work on. He had a pretty good corral and chute to catch and hold them.

The first two went well, no problems. The next had one normal testicle, but the other was high along the belly wall. I finally got both removed. From there, it just got worse and worse. Some calves had two normal testicles, but most did not. Others were along the belly wall or still in the belly where they originate during pregnancy.

It was a disaster. I could only do a few successfully. Many remained bulls or partial bulls.

Although I was frustrated, I now knew why he got bad prices for his bull calves.

"Do you have a new bull?" I asked, thinking that could be the problem.

"No," he answered, "I've had the same bull for years. I keep him because he's easy to be around. Never causes any trouble."

I needed to see his bull. This was confusing.

He led me to a different corral where the bull and cows were together.

I entered and walked through the cows until I found the bull. And yes, he was very gentle. Talking quietly, I could walk right up to him. He was unbothered. He just stood there.

I walked behind him. He remained quiet. I then was able to observe his scrotum and testicles. Or, rather, I should say his testicle. He only had one. Well, actually he had two, but the other was somewhere else. Probably still in his abdomen.

These things certainly happened out in the world of

animals. It was happening here at a high frequency. It is inherited and strongly so. The bull was passing this trait on to his male offspring.

The auction was never going to give him good prices until he got a new bull. A new bull that produced normal bull calves.

51

———

L

ubricating Joints, Please Don't

CLYDE HAD AN OLD HORSE. Actually, Clyde was old too. An old, beat-up cowboy who always had to have a horse. This was probably his last horse.

Rex was of working horse variety. Probably mainly Quarter Horse. He was old and had a big knobby knee. The left one. Lots of arthritis that had taken years to develop to this point.

Horses with big arthritic knees can get along pretty well for a long time. It's sort of surprising. Especially if you look at them. But if all you are doing is walking along the beach or maybe an easy trail in the mountains, they're fine.

But sooner or later, the pain gets to be too much, and you can't ride the horse anymore. All you can do is try to keep it comfortable. Drugs, given various ways, ointments. Things like that.

Or you can get really desperate and try the latest rumor or even an old wives' tale.

Clyde wanted me to look over his old horse, Rex.

He was really lame, I discovered. His leg was swollen, painful and hot. It really hurt him. I examined it carefully and soon realized the skin was damaged. It was swollen and painful to the touch. Something had really injured it. It seemed like a burn, a chemical burn. I actually thought I could smell a slight chemical odor, but I wasn't sure.

I told Clyde what I thought and asked what he thought might have happened.

He was reluctant to talk about it. Then I told him I really needed to know so I could help his horse. His horse, though old and crippled, was very important to him.

Finally, he told me. I guess it was more like a confession at this point.

He had sprayed WD-40 on it. Lots of it. A friend had suggested it. The friend said people sprayed it on their arthritic hands and fingers.

If you don't know about WD-40, it's a petroleum-based lubricant. It comes in a spray can. Quite honestly, it's a wonderful product for lots of things. Squeaky hinges, rust, stuck nuts, stuck screws, on and on. But not joints, man or animal.

Anyway, petroleum products are full of toxins.

I wasn't sure how that would work except, maybe, it caused enough swelling and pain that you forgot about your arthritis.

At least I knew what had happened and now could help his horse.

Still, I was curious about using WD-40 in people and animals with arthritis. After some searching, I found the company was located in San Diego. I found a phone number too. So, I called.

I explained my findings with Rex and that the owner had used WD-40.

I don't know if this is really possible, but I was sure I could hear their eyes roll when I spoke.

As it turns out, they were used to hearing about this dilemma. They were also tired of hearing about it. They strongly disclaimed any value in using their product for joint arthritis. They wanted it to stop. Which is, of course, understandable. It would only be a matter of time until a serious disaster occurred.

So, if there is a moral to this story, it is this: Don't spray WD-40 on your aching joints or your horse's aching joints.

52

L oose Wheel

HAVE you ever had something like this happen? Something so fast. Maybe terribly dangerous. Maybe even fatal. But it had happened so fast that you couldn't react. And then suddenly, it's over. Just like that. Like nothing ever happened. And everything is back to normal.

Of course, you only get to write about those things if nothing really does happen. Terrible anyway.

That doesn't mean you ever forget it. Never.

We were returning home on a Sunday afternoon. A nice fall day. Backcountry road in the big valley. Straight for miles. It was easy to drive 60 miles per hour. And seemed perfectly safe. So, we were driving 60 miles per hour.

No traffic either. Then we finally came up to an old car traveling our direction. Spewing exhaust, dented, rusty and driving much slower.

After following it for a bit, I decided to pass. There were no oncoming cars. It should be easy and certainly not dangerous.

I pulled out into the other lane to pass. As I got into the other lane, the front left wheel came off the older car. It moved into my lane, coming straight at us and bouncing.

It happened so quickly, I had no time to react. No time. I was just a spectator. The loose wheel was going to hit us. Most likely on the passenger side.

Then suddenly as it got to our car, it took a big bounce. A really big bounce. And it bounced over us. Completely. It completely missed our car.

I just kept driving. It was as though nothing had happened. As nothing frightening or dangerous had happened.

At times over the years, I have thought about it. Many times, actually. The wheel coming toward our car, then unexplainably bouncing over us. Completely missing us.

No tragedy. Just a haunting memory of what could have happened.

ABOUT THE AUTHOR

Russel Hunter DVM was born and raised in Southern California. He attended veterinary school at the University of California at Davis, and graduated in 1965.

He practiced as a country veterinarian in the San Joaquin Valley and the North Coast in California, before settling near Astoria, Oregon, for the next 50 years.

When not practicing, he traveled widely to many parts of the world. Dr. Hunter climbed mountains, kayaked Arctic rivers, backpacked extensively, and assisted with wildlife research in South America.